ArcView® GIS/ Avenue™ Developer's Guide

Amir H. Razavi

ArcView® GIS/Avenue™ Developer's Guide

By Amir H. Razavi

Published by:

OnWord Press
2530 Camino Entrada
Santa Fe, NM 87505-4835 USA

Copyright © Amir H. Razavi
Second Edition, 1997
SAN 694-0269

10 9 8 7 6 5 4 3 2 1

Printed in the United States of America

Library of Congress Cataloging-in-Publication Data

Razavi, Amir H., 1957-
   ArcView/Avenue developer's guide / Amir H. Razavi. –2nd ed.
      p.   cm.
   Includes index.
   ISBN 1-56690-118-9
   1. Computer software—Development. 2. ArcView. 3. Avenue
(computer program language).   Title.
QA76.76.D47R388 1997
910'.285'574—dc21   96-49394
                                                              CIP

Trademarks

All terms mentioned in this book that are known to be trademarks or service marks have been appropriately capitalized. OnWord Press cannot attest to the accuracy of this information. Use of a term in this book should not be regarded as affecting the validity of any trademark or service mark.

ArcView and ARC/INFO are registered trademarks of the Environmental Systems Research Institute (ESRI), Inc., the world's leading supplier of geographic information systems (GIS) software. Avenue is an ESRI trademark. OnWord Press is a registered trademark of High Mountain Press, Inc.

Warning and Disclaimer

This book is designed to provide information on customizing the ArcView program through Avenue, ESRI's object oriented programming language for ArcView. Every effort has been made to make the book as complete, accurate, and up to date as possible; however, no warranty or fitness is implied.

The information is provided on an "as is" basis. The author and OnWord Press shall have neither liability nor responsibility to any person or entity with respect to any loss or damages in connection with or arising from the information contained in this book.

About the Author

Amir H. Razavi is a professional engineer registered in the states of Maryland and Virginia. He has a B.A. degree in civil engineering, and a Master's of information systems management from George Washington University. Amir has been developing software since 1982, and has served as GIS manager for the Civil Rights Division at the Department of Justice. In 1994, he founded Razavi Application Developers (RAD), which specializes in ArcView, Oracle, and the Internet, among other applications. He has also co-authored *The ArcView/Avenue Programmer's Reference* (OnWord Press, 1995). The author can be reached on Compuserve at *71764,3262* and *71764.3262@compuserve.com* on the Internet. See RAD's home page at *http://ourworld.compuserve.com/homepages/razavi.*

Acknowledgments

This book would not have been a reality without the efforts and guidance of my editors, Barbara Kohl and Leslie Wagner, and my project manager, David Talbott, at High Mountain Press. I am very thankful to all three. I also wish to thank the Baltimore County government for permission to use the application presented in Chapter 15.

I have been blessed with a wonderful family that has supported me in every endeavor. Thus, I am dedicating this book to everyone in my family: to my wife, Tema, who inspired me to write this book; to my son, Dean, who cheered me along the way; to my parents, who have dedicated their lives to their children; and to my sisters, Homeria and Marva.

OnWord Press...

Dan Raker, President and Publisher

David Talbott, Acquisitions Director

Barbara Kohl, Senior Editor and Book Editorial Manager

Carol Leyba, Senior Production Manager

Michelle Mann, Production Editor

Cynthia Welch, Production Editor

Leslie Wagner, Assistant Editor

Daril Bentley, Senior Editor

Lisa Levine, Senior Editor

Kristie Reilly, Assistant Editor

Lynne Egensteiner, Cover Designer, Illustrator

Contents

Chapter 3   Avenue Programming Language   45

Chapter 4   Customizing the Interface   91

Chapter 5   Programming the Project Window   143

Chapter 6   Programming View documents   173

Chapter 7   **Programming Themes**   **187**

Chapter 8   **Programming Table Documents**   **201**

Chapter 9   **Accessing and Editing Databases**   **217**

Chapter 10   **Programming Chart Documents**   **235**

Chapter 11   **Programming Layout Documents**   **251**

Introduction

ArcView version 3, developed and distributed by the Environmental Systems Research Institute (ESRI), is a new approach to spatial analysis. Prior to the advent of ArcView, exploring spatial data involved extensive training and often extensive programming. Consequently, technicians usually performed analysis or developed customized data query and analysis routines for business users. By providing an intuitive graphical user interface, ArcView goes a long way toward bringing non-technical users closer to geographic information systems (GIS).

Similar to many other commercial software packages, ArcView is designed to appeal to a broad range of users. As a result, sooner or later you will reach a point where customization of ArcView is desirable to fit your unique needs. By using Avenue, ArcView's programming language, you can customize the program and further extend its power.

What Is Avenue?

Avenue is an object oriented programming (OOP) language, the most recent advance in software development technology. With Avenue you can create a new interface for ArcView or customize the current one, automate repetitive tasks, and write complete query and analysis applications. Avenue is equipped with a library of classes that represent the objects found in ArcView. The program executes tasks by accessing and manipulating these objects.

What Is Object Oriented Programming?

A programming paradigm is the basic approach to writing a program, and different paradigms satisfy different types of software applications. The procedural paradigm is well known. Other paradigm types include objected oriented, rule based, and visual.

One way to describe object oriented programming is through metaphor. Computer science is replete with metaphors, such as "memory" and "windows." Although what metaphors represent in computer science is far from what they resemble in the physical world, they are useful for understanding software functionality. The basic metaphors for object oriented programming are "objects," "classes," and "requests."

An object consists of both code and data, and can be conceptualized as "smart code" or "active data." A class is the blueprint that creates a functioning object, but is also an object with very limited capabilities. An analogy here would be that while a class provides the genetic code for creating hens, the actual hen object, not its class, lays eggs. Next, objects communicate, such as in an object receiving a request, or objects exchanging requests with each other. Requests ask objects to perform tasks, and tasks are the functionalities embedded in the object.

Object oriented languages are categorized as "pure" or "hybrid" systems. In the case of C++, a hybrid object oriented language, objects coexist within a procedural programming language. In pure object oriented languages, such as Smalltalk and Avenue, everything is an object. While Avenue does not implement all object oriented programming techniques, the current version is impressively powerful and can satisfy nearly all analysis applications.

Why Object Oriented Programming for Avenue?

The object oriented paradigm has addressed many problems inherent in the procedural approach to programming. The principal advantage of the object oriented language is its ability to handle complexity in a transparent manner. Under the object oriented paradigm, functionalities are placed inside the objects and out of the programmer's sight. Consequently, to execute a specific functionality, the programmer simply makes a request to the object.

When you become familiar with Avenue and the object oriented paradigm, you may notice that this paradigm eliminates redundant code. You may also notice that you save time by building a program with objects, and that the quality of your programs has improved because object functions are protected from any code you write. Finally, you may become convinced that the object oriented approach is a superior way to program.

How This Book Is Structured

Chapters 1 through 4 introduce the concepts of application development, ArcView interface customization, and Avenue as an object oriented programming language. Chapters 5 through 11 are dedicated to how the diverse parts of ArcView are accessed and programmed using Avenue. Chapters 12 through 14 focus on advanced topics and housekeeping. Chapter 15 lists the code for a fully developed application.

There are four appendices. Appendix A contains the Avenue class hierarchy. Appendix B provides a list of words reserved for use by the Avenue program. Appendix C lists the changes required to Avenue scripts created with ArcView version 2.1. Appendix D suggests programming guidelines for Avenue scripts. Finally, there is a complete index at the end of the book.

How To Read This Book

Most people involved in developing GIS applications are professional geographers, cartographers, surveyors, planners, geologists, and so forth. If you are among the ranks of these professionals, then you know that there is barely enough time to keep up with the technological advances in your own field, much less software engineering. However, if you have written programs in procedural languages such as Basic, COBOL, C, or AML (ESRI's Arc Macro Language used in ARC/INFO), you will know that object oriented languages constitute a new and different approach to programming.

This book attempts to accelerate the process of learning Avenue. After describing the basics of object oriented programming, the book then demonstrates how object oriented programming is implemented in ArcView and how you can take advantage of it.

You do not have to be a seasoned programmer to understand the *ArcView/Avenue Developer's Guide*. However, this book is about programming with Avenue, and is not a programming tutorial. Next, you do not have to be a GIS expert to benefit from the guide, but effective application developers understand the target environment. If you plan to extend the power of ArcView beyond what you are able to do with the package as shipped from ESRI, this book is for you.

Because this book is about application development, a discussion of the application development process is included. Chapter 1 reviews the structured approach to application development. If you are a seasoned software engineer, this chapter can serve as a quick review.

If you are not familiar with application development, read Chapter 1 in order to learn about the recommended method of application development. Countless volumes about development approaches are available, and the few pages appearing here barely scratch the surface. The purpose of this chapter is to whet your appetite for additional readings on the subject.

If you are new to object oriented programming languages, read Chapter 2, "Avenue Fundamentals," very carefully. The chapter describes the basics of object oriented programming and how they are implemented in Avenue.

Chapter 3, "Avenue Programming Language," presents Avenue's programming elements. You must read this chapter if you are new to programming. Experienced programmers should review this chapter to become familiar with the Avenue language.

Avenue scripts are implemented through customized controls such as menu items and buttons. Chapter 4, "Customizing the Interface," discusses these controls and explains how they are merged with Avenue scripts. For both experienced and fledgling programmers, this chapter provides a preview of the task environment that you wish to modify through Avenue.

Chapters 5 through 11 demonstrate how various elements of ArcView are accessed and programmed using Avenue. The information in these chapters will be applied in most of your application programs.

If you plan to distribute your customized ArcView application, you should study Chapter 12. This chapter discusses how to protect your application when installed at user sites.

Chapters 13 and 14 are advanced topics that you may want to read after you have gained some experience in programming with Avenue. Address matching with Avenue is explained in Chapter 13, and Chapter 14 focuses on integration of ArcView into other applications.

Chapter 15 presents and describes eight Avenue scripts for a sample application.

Typographical Conventions

⊷ **NOTE:** *Information on Avenue features and procedures that are not straightforward or intuitive, may appear in a note.*

✓ **TIP:** *Tips are the fruit of experience, and are aimed at saving you time and stress.*

Avenue Scripts following this symbol are available on the attached diskette.

Text (code) that you type into an Avenue script appears in a monospaced typeface as shown below.

```
if (nil = theView) then
```

❑ Menu items, tool buttons, and programming elements, such as requests and object classes, are capitalized.

The Category option must be set to Menu.

❏ Names for files, directories, fields, and user-derived input of diverse kinds appear in italics.

Avenue scripts in this chapter use the *USA* coverage bundled with the ArcView software. The coverage is located in ArcView's *AVDATA* directory.

Companion Disk

The program scripts in this book highlighted with a disk icon are stored on the companion disk for personal reference. The disk contains 14 subdirectories, each named by chapter number. All the scripts within that chapter appear in individual files. For example, the second script in Chapter 4 will be found in the file named *04-02.ave*. This companion disk does not contain ArcView software. ArcView software is necessary in order to use this material.

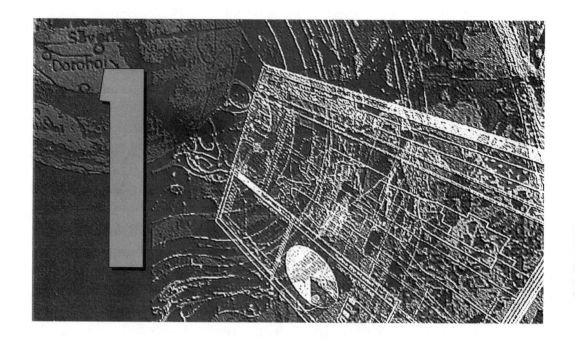

Structured Application Development

Similar to computer applications in other fields, GIS applications must be developed according to a structured methodology. If you plan to write short Avenue scripts to automate a few tasks, development methodology may not be for you. If you plan to develop complete applications or implement new systems, following a structured methodology dramatically increases the probability of a successful product.

For many people "methodology" implies extensive overhead, extraneous tasks, and the stifling of creativity. In reality, however, adopting a methodology simply means using certain techniques and executing certain tasks that will produce the best results.

You may be familiar with one of several development methodologies tested over the past several years, particularly if you have developed ARC/INFO applications. The methodology discussed below is both a combination of and variation from methods used in developing GUIs (graphical user interfaces) and client/server application software. It is also somewhat different from traditional methods used in developing ARC/INFO applications because ArcView is a new generation product.

The proposed structured methodology for developing ArcView applications is comprised of the following stages.

❑ Requirement study

❑ Prototyping

❑ Construction

❑ Structured testing

In the remainder of this chapter, the four stages are described and illustrated through a hypothetical scenario. This scenario assumes that we are GIS consultants to a large metropolitan bank. Our task is to develop an application showing characteristics of mortgage applicants and recipients within designated geographic areas.

Because the techniques and procedures for each stage in the recommended methodology have been applied in developing many software applications, detailed information is available in numerous publications on software development methodologies. See the concluding section of this chapter for a list of publications. Next, it should be mentioned that development methodology is only one part of implementing a GIS project. Other issues such as hardware, data sets, staffing, and procedures must also be considered, but are beyond the scope of this book.

Requirement Study

The requirement study stage begins upon recognizing that a solution is required for a particular problem. Its purpose is to develop a specification document describing what the software will accomplish, but without explaining how the software will work.

The specification document is the bedrock of your application. The more accurate and complete the foundation, the better your final product will be. Managers and analysts often ignore this stage in favor of writing program code because they prefer seeing a tangible outcome as early as possible. Nevertheless, experienced software developers would assert that you are doomed to fail if you ignore a study of software requirements.

Two distinct activities occur during the requirement study phase: problem analysis and product description. These activities are not carried out in serial fashion, but rather product description evolves as problem analysis progresses.

Because the purpose of analysis is to acquire a complete understanding of the problem, most of the analyst's time is spent in meetings with people who are knowledgeable about the problem. In the course of product description, the analyst develops the specification document that explains exactly what a software application can do to resolve the now clearly defined problem.

First Faire Bank's Requirements

During a meeting with First Faire Bank (FFB) managers, we learn that the bank president Donald Zinger is concerned about recent media attention on alleged discriminatory lending practices by other banks in surrounding counties. Since FFB management is genuinely committed to a nondiscriminatory lending policy, Zinger wants to ensure that no one will be able to identify patterns that even remotely suggest discrimination. The president is aware that GIS can be used for monitoring any number of variables within specific geographic areas.

The requirement phase begins by identifying the type of evidence an adversary could use when claiming that FFB engages in discriminatory lending practices. FFB's chief legal counsel Mary Lawsome explains that in some cases, branch banks allegedly approved disproportionately fewer mortgage contracts in areas of high minority population density than in low density areas. In these instances, average household income was similar in all locales studied.

Next, we learn that annual reports on mortgage applicants generated by FFB include ethnicity and a variety of socioeconomic status variables. Applicants are also geographically referenced by census tract. In addition, the bank produces quarterly statistical reports on the distribution of active mortgages by ethnic group. However, because most bank managers consider these reports to be complicated, they do not bother to study them.

At this juncture, we decide that FFB needs a GIS application that presents branch offices, mortgage counts, and mortgage values over a map displaying minority population density. Required data components include census tract boundaries, demographic data, branch locations, and mortgages (applicants, approved contracts, and dollar amounts). Equipped with these data sets, the application could produce demographic maps, branch location maps, pie charts of mortgage numbers and amounts by census tract, or any combination of the above. We then document the requirement study and distribute copies for review to the president, general manager, and chief counsel.

Prototyping

Under traditional methodologies, analysts would write the design specification document following the requirement study phase. Next, users would be requested to approve the specification document before analysts commenced producing program code. However, because users could rarely fully comprehend the paper specifications, user approval at this phase in the process was generally meaningless.

The purpose of prototyping is to quickly develop a rough version of the desired system. Users can more easily comprehend the application by reviewing the prototype than by studying paper specifications. A prototype can also be adjusted and modified to show various design alternatives.

In ArcView, prototyping begins with customization of the interface controls and project components. Avoid writing Avenue scripts for the application logic as part of the prototyping. Instead, identify the required scripts and write a short description for each, and define and diagram the data flow. Review your prototype with users and describe how the application will behave once the scripts are developed and incorporated. Repeat this process until users are satisfied with your description of the desired system.

As you become more experienced with ArcView, Avenue, and this methodology, user participation in developing the prototype can accelerate the process. In fact, if you have strong user participation, you could combine the requirement phase with prototyping for simple applications.

Prototyping at the First Faire Bank

Within a few days of submitting the requirement document, FFB managers contact us with their responses. Fortunately, everyone likes the proposed GIS application. Because the legal office is to be the primary user, chief counsel Lawsome assigns a member of her staff, Richard Atty, to work with us in developing the application. We also request that the general manager assign a financial analyst to assist us in deciphering the annual mortgage reports.

A quick survey of the bank's legal office staff reveals that while everyone uses personal computers and GUI applications, no one is familiar with the GIS concept. Consequently, we present a half-day seminar to the legal office staff. This consists of an introduction to GIS and a demonstration of ArcView. As a courtesy, we invite the bank president and general manager, who accept and attend the seminar in the company of several other management staff members.

After the seminar, we spend a few hours with Atty in order to demonstrate selected ArcView applications in detail. We explain how the software can be customized to suit the bank's needs.

We then decide to begin prototype design by customizing the interface and preparing sample data. After the first cut of the prototype is complete, we rehearse a scenario to take users through the application. Next, we meet with Atty to review the prototype. He is very excited about our work, but makes no critical comments or suggestions. Either we have done a terrific job, or the user is not prepared to participate. We decide to be cautious, and give Atty a few extra days to digest the prototype. We make it easy for him to access the prototype, and we remain available to answer questions by phone.

We stay in touch with Atty to make sure he is experimenting with the prototype and has not forgotten us. Within several days, Atty breaks his code of silence and starts asking questions. Questions eventually lead to suggestions and recommendations. As a result, we prepare a second version of the prototype and demonstrate it for all legal office staff. The demonstration generates a few comments on required and desirable functionalities. After we have incorporated the results of all commentary into the third version of the prototype, the legal office staff agrees to accept the prototype. At this point, we print the customized interface and document the functionality that each interface control must provide.

Construction

In the construction phase, the Avenue scripts required for the application are developed. Standards for current and future applications should be established before writing code. The extent and range of the standards depend on your application. Script documentation and naming conventions are minimum standards. You should also determine how common tasks such as error handling and data sharing procedures will be accomplished.

Similar to many GUI applications, ArcView applications consist of numerous scripts that have no apparent relation to each other. The scripts are usually linked to a control interface and perform their work independently. Before writing each script, document in plain English how the tasks are carried out. This practice will help you develop clean and efficient scripts. Keep in mind that because Avenue is not a procedural language, ArcView does not respond to requests in a serial fashion.

Applications written with procedural programming languages control user access to features or functions. In contrast, the users control access to features or functions in ArcView applications written with Avenue, and in other GUI applications.

For example, a GIS application could show sewer pipes installed prior to a certain year within a specified zip code area. In ARC/INFO, such an application is written with AML, a procedural language. The ARC/INFO application asks for the zip code before zooming into the area. The user is then asked to input the year before the application displays the pipes.

The same application in ArcView would present menu items and buttons to perform each task independently of other tasks. In short, the user controls the order of the process. For instance, the user could choose to display all the pipes first, and then provide a year to select from the displayed pipes. Or the user can choose to zoom in to a particular zip code area.

In AML, tasks are carried out serially, which means that one task must be performed before the next in the series can begin. In Avenue, tasks are carried out by sending a request to an object. (Requests and objects are discussed in detail in Chapter 2.) Note that the object receiving the request will carry out the task. Depending on the request, Avenue may not wait for the object to finish its task before moving on to the next task.

For example, when an Avenue script designed to change the map scale iterates through a loop five times, the map scale is changed every time. However, the map display you see is the result of only the final scale change. Every time the scale is changed, a request is sent to update the map with the new scale. Avenue does not wait for the map on the screen to be updated; the scale is changed and another request is sent that overwrites the previous one.

Construction of the First Faire Bank Application

While waiting on user reviews during the prototype phase, we prepare for the construction phase. Because the bank has no previous experience with GIS applications, specific GIS-related standards do not exist. However, FFB programmers have established GUI standards for their Windows applications. By the close of the prototyping phase, we have reviewed the bank's GUI standards along with the file structure of the required data as it is maintained on the bank's mainframe. We have also added our standards for error handling and naming conventions.

We are now ready to create Avenue scripts. The document we generated at the end of the prototyping phase can organize our progress through the construction phase. We will check off each functionality as we complete respective Avenue scripts. Involvement with the users at this stage is limited to clarifying or resolving incidental issues.

Structured Testing

The purpose of testing is to identify application software defects. Resolving as many defects as possible before delivery is important. People will lose faith in your ability as a developer if your software product is riddled with defects. Therefore, testing must be viewed as a destructive process in the search for software errors.

There are three levels of software testing. "Unit testing" is executed by the programmer to ensure that each Avenue script behaves as expected. "System testing" is also carried out by the programmer, except in cases involving large projects where an independent test team accepts the responsibility. The purpose of system testing is to determine whether various scripts work together properly and whether the system meets requirement specifications. "User acceptance testing" is executed by users to determine if the application meets their needs.

Structured Testing of the First Faire Bank Application

Because we are seasoned developers, we do not succumb to unstructured testing, a common trap for many novice programmers. Unstructured testing tends to produce favorable results because programmers unconsciously test in a manner that avoids software failure.

Instead, we begin by developing a plan for testing each script and the entire application for boundary data, typical data, and complete code coverage. Boundary data refers to cases in which data values extend beyond what is considered normal or acceptable. Complete code coverage refers to executing the script under various conditions to ensure that each line of script is executed at least once.

Subsequently, we develop test scripts for each unit and the entire application. Each script describes the input, expected output, and procedures. Next, we carry out each test script, record the output, and compare it with the expected output.

A few inconsistencies are investigated and corrected. Upon submitting the application to the FFB legal office staff for testing, we are informed that in a week it will be officially drafted into use, provided we change three minor interface features.

The First Faire Bank scenario was perhaps a bit smoother than most structured application developments in the real world. However, the stages described here constitute the bedrock of effective application development, and cannot be overemphasized.

Further Reading

You can find more information about application development in the following publications.

James Rumbaugh. *Object-Oriented Modeling and Design.* Prentice-Hall Inc., 1991.

James Martin. *Rapid Application Development.* Macmillan Publishing Co., 1991.

Alan M. Davis. *Software Requirements: Analysis and Specification.* Prentice-Hall Inc., 1990.

Sanjit Mitra, ed. *Handbook of Software Engineering.* Van Nostrand Reinhold Co., 1984.

Tom Demarco. *Structured Analysis and System Specification.* Prentice-Hall Inc., 1979.

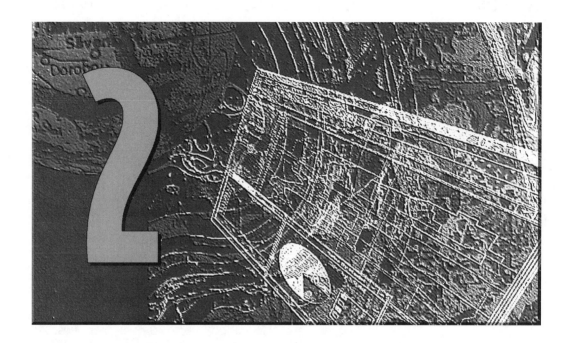

Avenue Fundamentals

Welcome to Avenue. In this chapter, Avenue's basic characteristics and building blocks are presented, followed by a discussion of how to prepare and run Avenue scripts. You will soon discover that you can extend the flexibility and power of ArcView by using Avenue.

Avenue Building Blocks

Class is the most basic concept in an object oriented programming language like Avenue. A class provides the blueprint for creating objects, and is itself a type of object. Programs communicate with objects by passing messages. These are known as *requests* in Avenue. If the request is recognized by the object, it provides a predefined service and optionally returns a resultant object. Objects, classes, and requests are the basic building blocks of Avenue.

Objects

An object is defined as a person, place, or thing that knows the values of its attributes and can perform operations on those attributes. It models the real world. Objects are also known as *class instances*, and operations are also called "methods" or "services." In ArcView, objects are the things that you work with, such as a display extent. A display extent in ArcView is defined as a rectangle. Attributes for a rectangle are the coordinates of the corners. The rectangle could provide a service that would change its size and location.

Classes

A class groups objects based on common characteristics. For instance, a display extent object belongs to the rectangle class, and an individual person belongs to the human class. All objects of a class share the same attribute types and operations. For instance, all individuals have a birth date because an attribute of the human class is to have a birth date.

There are two types of classes: *abstract* and *concrete*. Abstract classes do not instantiate objects, while their descendants have objects. The purpose of an abstract class is to model common attributes and operations of several classes into a single class. When a class can instantiate objects, it becomes a concrete class. For example, consider the class Doc, which has no objects; its descendant classes, such as View and Chart, do have objects. Therefore, Doc is an abstract class while View and Chart are concrete classes.

Requests

In order to invoke an operation within an object, you must send a request to that object. Similarly, you must first ask a person how old they are in order to know their age. The process of asking, which is the request, starts the calculation operation in one's mind and results in a reply, which is the age. A request may require parameters in order to perform the corresponding operation.

For instance, a request to zoom into a display area requires a percentage value in order to determine the degree of enlargement. The request is carried out by the object if the request has a corresponding service of the same name, and the parameters required by the service are specified. Requests may also be made to a class, but such requests are generally limited to creating class instances.

Avenue Characteristics

Avenue is ArcView's scripting language, and it can be used in the same fashion that AML (Arc Macro Language) is employed to create ARC/INFO applications. Avenue and AML are based on different development roots. While AML is a procedural language, Avenue is an object oriented programming language. Object oriented languages provide a higher degree of uniformity, stability, and reusability than procedural languages because of the three major characteristics defined below.

Encapsulation = to make concise, condense

This characteristic is also known as "information hiding." Procedural programming languages maintain data separately from procedures. In object oriented languages, data and procedures are kept together. An object consists of both data and code. Therefore, the paradigm for performing operations in an object oriented language is sending a "request" to an object that tells the object what you would like it to do. Requests are similar to function calls in procedural languages. The benefit of encapsulation is that it reduces interdependencies among application modules.

par´ə dim
"an example"
" or model"

For example, an application may perform zooming in to a map display by sending a request with a rectangle object as its parameter. The algorithm used for zooming is hidden from the application and it may change from one version of ArcView to another. However, as long as the name and structure of the request do not change, changes to the zoom method have no effect on the application. Hence, the application has gained some degrees of stability.

Polymorphism

Polymorphism allows different objects to respond to the same message in their own unique ways. In most procedural languages, it is necessary to name functions differently when data types are different. For instance, adding two numbers may require a function named *Add_Numbers* while concatenating two strings requires another function named *Add_Strings.*

In an object oriented language, only *Add_Objects* is defined, and the compiler selects the correct service based on the data type. Under polymorphism, the request will always be Add_Objects, even though the code that provides this service may vary from class to class. This behavior provides a uniform interface within the program because the same name is used, while differences in code remain invisible to the programmer.

Inheritance = Receiving

The inheritance characteristic of object oriented languages allows the reuse of software by specializing already existing general solutions. Attributes and operations can be shared among objects in a hierarchical relationship. Items are inherited by the children from parents through the hierarchy. For instance, the classes Rectangle and Circle are children of the Shape class. Any attribute or operation defined for Shape is available to Rectangle and Circle. The parent class is also called the "superclass." When a request is sent to an object, the compiler searches for the corresponding service. If the compiler finds the corresponding service, the request is performed. Otherwise, the search process is moved to the object's superclass. This process continues to the top of the class hierarchy. Avenue produces an error message if it cannot find the request.

Interaction Between Building Blocks

Avenue programs are an organized collection of requests made to classes and objects. Applications developed in Avenue accomplish tasks by accessing objects and sending requests to objects. For instance, an application can zoom in to a map display by sending a zoom request to the display. The request process is the interaction between Avenue building blocks.

Creating and Accessing

To create an object, you must know its class. ArcView's class hierarchy, accessed through the help system, provides an efficient method for searching classes. Once you have become proficient in ArcView class hierarchy and definitions, it should be used as a verification tool. (See Appendix A for diagrams of the class hierarchy.) When you create an object, you need a variable to reference it. ArcView maintains each object in a memory area and the variable points to that memory location. Chapter 3 explains variables in more detail. An example of creating a new object follows.

```
MyView = View.Make
```

In the preceding statement, the Make request is applied to the View class to create a new view object referenced by the MyView variable.

You may need to use an object that already exists but has no variables to reference it. For instance, assume your application needs to zoom in to the display area of an existing view. You do not need to create a new view object or a new display object. Instead, you want to access the existing display object so that you can send it the ZoomIn request. In such cases you start with an object that you can access and use the association between the objects to reach the desired one. This method is demonstrated in the following code segment.

CODE   02-01.AVE

```
myProject = av.GetProject
myView = myProject.FindDoc ("Atlanta")
myViewDisplay = myView.GetDisplay
myViewDisplay.ZoomIn (150)
```

In the preceding example the objective is to zoom in 150 percent to the display area of an existing view. The display object is associated to a view object. Your view object is associated to a project object which is associated to the ArcView program. The object that represents the ArcView program is always known to you. Therefore, starting with the object that models your ArcView program you can traverse back through this chain of associations to access the display object.

The object that models your ArcView program is instantiated from the Application class. It is referenced by the keyword *av*, and is the sole instance of its class. In the last example, sending the request GetProject to the object *av* returns the project object. You would then send the FindDoc request to the project object in order to access the view object. The last object to reach is the display object for the view. When you send the GetDisplay request to the view object, you get the display object. Finally you can send the ZoomIn request to the display object that is referenced through the myView Display variable.

The association between objects is very important, as demonstrated by the last example. The association is used to proceed from one object to another. In order to learn these associations you must understand ArcView's object model, which is described in the next section.

NOTE: *ArcView does not permit the creation of new classes or requests.*

In this book, the term *object* is often used when referring to a variable because variables are used to reference objects.

Making a Request

In order to initiate an action from an object or class in ArcView, you must send a request to the object or class. Requests are sent in one of three ways: *postfix*, *infix*, and *prefix*. The three following examples demonstrate each method.

```
myView.CopyThemes ' postfix
tomorrow = toDay + oneDay ' infix
Not true ' prefix
```

The single quote character in Avenue represents a comment string, one of the language elements discussed in Chapter 3. In the first line, the postfix Copy-Themes request is applied to an object with a period. Most requests are applied using the postfix method. The infix example shows how the addition request sits between two objects. Finally, the prefix example places the not request before the object called True. Requests can be sent to both classes and objects.

Requests sent to objects are known as *instance requests*, while requests sent to classes are called *class requests*. ArcView's on-line help shows valid requests for an object or class and how they are applied.

Requests can carry parameters required by the object or class. As shown in the following example, the parameters are passed along with the request by placing them in parentheses after the request.

```
MsgBox.Error(aMsg,aTitle)
```

The above Error request is asking the MsgBox class to create a window with an OK button and Stop icon. It also requests that the contents of the object *aMsg* be used as the error message, and that the contents of the object *aTitle* be placed in the window's title bar.

The ArcView on-line help can show details of a request's parameters. You must provide the same number and type of parameters in the order identified by the ArcView's documentation. In the example above, if you do not provide a title, ArcView uses the word "Error." This does not mean that only one parameter is passed to the object, but rather that the object *aTitle* does not have a string value. Use double quotation marks, as shown in the statement below, to indicate an empty string. The following illustration shows the result of the statement.

```
MsgBox.Error("Unable to open theme", "")
```

Example of an Error request to the MsgBox class.

Request names follow a convention that represents action and property. Learning and understanding this convention make you more efficient in identifying the request you need to use and its expected result. The convention is based on having a request name that starts with one of the following words: *add, as, can, find, get, has, is, make, return,* or *set.*

Requests beginning with the word *add* add or aggregate one object into another. For instance, a view object may contain themes. In order to add a theme object to a view you would use the AddTheme request. As shown in the following example, a theme object referenced by *myTheme* variable is added to the *myView* object.

```
myView.AddTheme (myTheme)
```

Requests starting with the word *As* convert the objects of one class to objects of another class. A common use of these requests is to convert non-string objects into string objects in order to display or write them to a file. When an AsString request is sent to an object, it returns its string representation. An example of this request in shown in the following code segment.

 02-02.AVE

```
today = Date.Now
nextYear = today + (1.AsYears)
MsgBox.Info ("Same time next year is"+NL+
nextYear.AsString,"")
```

Requests that start with the word *Can* indicate the ability of an object or class to perform a task. These requests are useful in writing proactive applications. For example, if your application accepts a file name from the user to load into a table document, you should first check if a table can be created from that file. This is accomplished by sending the CanMake request to the VTab class. In ArcView, objects of the VTab class represent the tabular data shown in the table document. Another example follows:

02-03.AVE

```
myView = av.GetActiveDoc
myActiveTheme = myView.GetActiveThemes.Get(0)
canDoIt = myActiveTheme.CanLabel
if (canDoIt) then
    myView.LabelThemes (true)
else
    exit
end
```

You can find an object by using requests that start with the word *find*. The most common use for this request is to find aggregated objects by name; for instance, the FindDoc request finds ArcView documents in a project. Another example is the FindTheme request that finds a theme in a view. In the following code segment, FindStr is used to search the attribute fields of a theme for the word "Chester."

```
myView = myProject.FindDoc ("USA")
countyTheme = myView.FindTheme ("County")
countyCenterPoint = countyTheme.FindStr ("Chester", true)
```

When you have access to an object, you can obtain the value of its attributes or access other objects associated with it. Requests starting with the word *Get* return an attribute of an object or reference to an associated object. An example of getting an object's attributes follows.

```
rotationAngle = myNorthArrow.GetAngle
```

An example of referencing an associated object is accessing a view's display object, as shown below.

```
myViewDisplay = myView.GetDisplay
```

You can identify the state or contents of an object by using requests that start with the word *Has* or *Is*. For instance, you can verify if a theme has an associated table document by sending the HasTable request, or verify if a theme is active by sending the IsActive request. These requests return a True or False response.

Make requests are applied to classes to create new objects, as illustrated below.

```
myRectangle = Rect.MakeXY (0, 0, 100, 100)
newTheme = Theme.Make (mycoverage)
```

Requests starting with the word *Return* function similarly to requests starting with *Get*, with one major difference. The Get request provides access to the object while the Return request creates and provides access to the copy. Consequently, changing objects accessed through the Get request changes the original object, while changing objects accessed through the Return request has no effect on the original object. In the following example, *extentRect* is equivalent to the view's display extent, but changing it does not change the extent of the view.

```
extentRect = myViewDisplay.ReturnExtent
```

In order to establish or change an attribute of an object, you would use requests that start with the word *Set*. For example, to change a view's name you send a SetName request as shown below.

```
myView.SetName ("USA");
```

Requests may be chained for code efficiency. ArcView processes chained requests from left to right. Chaining requests can reduce memory load by not setting variables. Avoid extensive chaining because it makes your script harder to read and more prone to errors. The following code segment shows an example of chained requests.

```
av.GetProject.FindDoc("Virginia").FindTheme("Counties").
SetThreshold(aThreshold)
' The above statement is equivalent to the following code segment.
thisProject=av.GetProject
aView=thisProject.FindDoc("Virginia")
aTheme=aView.FindTheme("Counties")
aTheme.SetThreshold(aThreshold)
```

Avenue Statements

The interaction between the building blocks of Avenue is captured in Avenue statements. In its simplest form, an Avenue statement is comprised of a request and an object or class. An example is *myTheme.Set-Name ("County")*. Avenue statements have different forms. An assignment statement is formed when the request returns an object and that object is assigned to a variable, as shown below.

```
myFileName = FileName.Make ("report.txt")
```

Another type is the conditional statement, which returns a True or False. Conditional statements, also known as Boolean expressions, are described in detail in Chapter 3.

ArcView's Object Model

Developing ArcView applications requires familiarity with its object model. The more complex an application is, the more you need to understand ArcView's object model. It describes the objects, and their attributes, operations, and relationships to other objects.

ArcView's on-line help provides an adequate explanation of object attributes and operations. This section concentrates on the relationships among the objects. Earlier in this chapter you learned how to access ArcView objects by traversing associations from one object to another. The association of two objects describes their relationship. Therefore, it is important to learn these relationships for writing ArcView applications. ESRI has adopted the graphical notation used in object modeling technique (OMT) methodology to present these relationships. (See James Rumbaugh et al; *Object-Oriented Modeling and Design;* Prentice-Hall, 1991. This book also follows OMT diagramming techniques.)

Association implements the relationship among objects and classes and represents a logical connection between two objects or classes. For example, the application object is associated with a project object and the project object is associated with ArcView documents (e.g., each ArcView document is associated with a document window).

Associations can be single or bi-directional. In a bi-directional association you can start with either object and travel the association to access the other object. However, most associations in ArcView are singular in direction. For example, you can start from the project object and access any ArcView document by using the FindDoc request. You cannot, however, reach the project object from the ArcView documents since no request for it has been provided.

An example of bi-directional association in ArcView is the relationship between a document and its window. The document window object is the window from which a document is viewed. You can send the GetWin request to a document object to access its window. Conversely, you can send the GetOwner request to a window object to access its document.

Association in its simplest form is shown as a line between two objects or classes. The following figure shows how an association is sketched between two classes or objects. The association between two objects is also known as a *link*. A link is the instance of an association. The object diagrams displayed in this book show the association only between classes, which implies that association is also available for objects of the same class.

Association between classes and link between objects.

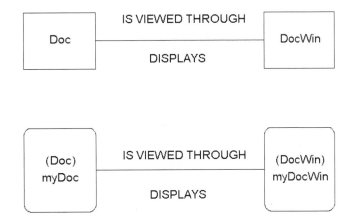

As shown in the above figure, an association may be named, and the name is typically a verb form. When the meaning of the relationship between two classes becomes obvious, then the association's name can be omitted. The name is tied to a direction. Rumbaugh recommends arranging classes in such a manner that the names are read from left to right, omitting the name for the reverse direction. ArcView's object model, both in ESRI's documentation and in this book, does not generally show association names because they can easily be identified by their classes.

Multiplicity

A straight line from one class to another indicates an association in which only one instance of each class participates in the relationship. This is known as a one-to-one multiplicity. Each Doc object is viewed through a single DocWin object. Multiplicity specifies how many instances of one class may relate to a single instance of another class. The following figure shows how multiplicity is sketched.

Multiplicity of associations.

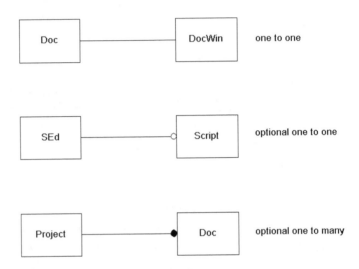

In a one-to-one relationship the association is mandatory—both objects must exist at the same time. In an optional one-to-one relationship, one object can exist without the other. This multiplicity is rare in the ArcView object model. In the preceding figure, the optional one-to-one relationship shows that a script editor and a script object can exist independent of one another. However, only one script can be related to a script editor.

The optional one-to-many relationship indicates that an instance of a class may relate to many instances of another class simultaneously. In this figure, you can see that a project object may optionally have any number of ArcView documents. Multiplicity may also be written on one end of the association line; for instance, "1+" can indicate one or more.

Aggregation

Association between objects can show aggregation, the "part-whole" relationship. In this relationship the component objects are aggregated into the assembly object. For example, a view document is composed of several themes. This relationship is shown in the following figure.

Aggregate relationship.

Aggregation is shown with a diamond next to the "whole" object. The association is usually named *contains* in the direction of the "part" object, and *is part of* in the reverse direction. Aggregation is not symmetrical, which means that if A ("part" object) is part of B ("whole" object) then B cannot be part of A.

Generalization

The inheritance characteristic of object oriented technology is modeled through generalization. Generalization is the relationship between a class and one or more of its specialized versions. The class that is being specialized is referred to as the "super-class" or "ancestor," while the specialized class is known as the "sub-class" or "descendant." For example, as shown in the following figure, ArcView's document object is specialized into view, table, chart, layout, and script editor documents.

Generalization.

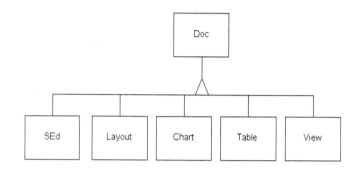

Generalization is shown with a triangle pointing to the super-class. This association is often named *is a kind of* in the direction of the super-class, and *has types of* in the reverse direction. The object model is used extensively in this book to explain the relationship among objects. Only segments of the model with relevant classes are shown for clarity.

Creating a Script

Avenue scripts are the code statements that collectively make up your application. If you are new to Avenue, start with simple scripts. Working out your script on paper first could make the process easier. You can then create your script by typing the code directly into the ArcView script editor, or by using your own editor.

As you learn Avenue, or if the scripts are small, you may find it easier to use ArcView's editor. However, once you start creating applications, using a system editor is advisable. There are two reasons for using a system editor instead of ArcView's. The primary reason is that ArcView stores scripts within the project file. By using a system editor you will have a text of your script to reuse. The second reason is that ArcView's editor has very limited editing features, and your system editor will probably be more efficient.

✓ **TIP:** *Having access to ArcView's help system while writing scripts is useful. Open the help system and iconize it during sessions with your editor.*

Using the ArcView Script Editor

The Script component provides an integrated development environment for developing Avenue programs. To open the component, select the Scripts icon from the project window. At this point you can create a new script or open an existing one. Default names for new scripts are *Script1*, *Script2*, and so on. Always change the default to a more meaningful name by selecting the Rename item from the Project menu option. You can also rename a script in the Script Properties dialog box.

The Avenue programming environment, shown in the
following figure, consists of the Script Editor window
and its associated menu options and buttons.

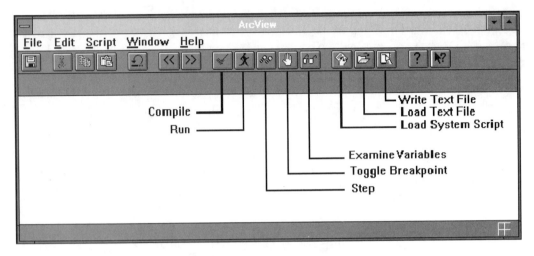

Avenue programming environment.

Scripts are saved within the project file rather than
independently. Therefore, once you have finished
writing a script, save it both as a script and a text file.
You can save your code as a text file by clicking on the
Write Text File button.

Using a System Editor

You can use any editor as long as the script is saved as
a text file. Once you have completed the script, open
a new Script Editor window by selecting the Script
icon from the Project window, and then click on the
Load Text File button.

A Load Script dialog box appears, as shown in the next illustration. Select the text file and click on the OK button. ArcView then loads a copy of the file into its script editor. The loading process does not involve compiling or error checking.

Load Script dialog box.

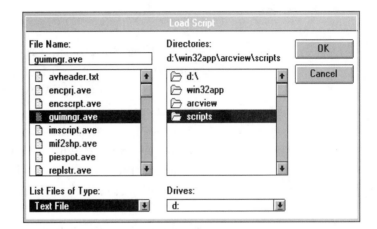

✓ **TIP:** *When you make minor changes in the Script Editor, do not forget to update your original text file. Update the original file by saving the script as a text file.*

Where to Start

When composing scripts, you need to access various objects to control or retrieve information. First, you need a starting object. For instance, to access a theme you first need the view object that is holding the theme. The most common method of retrieving the starting object is to find the active object. Assume that you have developed a script to hide the theme's legend, and you have associated this script with a button for view documents. When the button is pressed, the first thing that your script should do is retrieve the view object. In the following statement, the variable called *theView* refers to the active document, in this case a view.

```
theView = av.GetActiveDoc
```

In the above statement, a request is sent to the object called *av*. The *av* object is important because it represents the executable ArcView. There is only one instance of *av*, and that is your current application. Consequently, *av* is the starting object. Because your script is associated with a view control bar, the active document must be a view. The rest of the script will use the object called *theView* to access its themes.

⚭ NOTE: *An object is an instance of its respective class.*

The statement below is another common way to start a script.

```
thisProject = av.GetProject
```

The object *thisProject* represents the current project. This approach is often used when the script is not associated to a specific document, window, or control bar. With the project object you can add and remove documents, or save the project.

Accessing System Scripts

A system script can be accessed for every user interface control object in ArcView. The script that you are writing may be a specialization or combination of some of these system scripts. If you believe that a particular system script would be useful, go to the Customize dialog box and identify the script name. (See Chapter 4 on the access and use of the Customize dialog box.) Open a new script window and click on the Load System Script button to display the Script Manager. Scroll to the desired script, select it, and click on the OK button. The contents of the script are copied into your new script window. Repeat this process to insert other system scripts.

Script Testing

Avenue is a compiled language. Before using your script you must compile and test it for accuracy. While compiling and testing the script, Avenue also tests for syntax and run-time errors.

Script Compiling

Before you run an Avenue script, the code must be compiled. Click on the Compile button to check for syntax errors. If no errors are found, ArcView compiles your script and activates the Step and Run buttons.

Your script will not run if it contains syntax errors, which are caused by failure to follow programming language rules. Simple typing errors are often identified by the compiler as syntax errors. If the compiler detects a syntax error, it displays a message and moves the cursor to the corresponding location in the script.

The compiler locates one error at a time. Therefore, if you correct an error and recognize other similar errors in your script, you will save time by correcting them before compiling again.

Start the Script Editor and key in the following code segment as a new script:

```
' This code segment has a syntax error.
thisProject=av.GetProject 'starting object
' Accessing a view document.
aView=thisProject.FindDoc("states")
if(nil=aView)
   ' Was the view found?
   MsgBox.Error("States View not found","")
   exit
end
```

Compile the script. The error message shown in the following illustration appears and the cursor moves to the location of the error. Can you correct this error?

An error message.

The corrected code follows:

 02-04.AVE

```
' This code segment compiles without syntax error.
thisProject=av.GetProject 'starting object
' Accessing a view document.
aView=thisProject.FindDoc("states")
if(nil=aView) then '  Location of previous syntax
  'error.
  'Was the view found?
  MsgBox.Error("States View not found","")
  exit
end
```

Another common cause of syntax errors is the failure to define a variable or an object in the script. Avenue does not require declaration of variables at the start of a script. Variables are declared as references to objects when they are used on the left side of an assignment statement.

Running Your Script

You should test each script before associating it to a control object. Testing each piece of the application individually is known as "unit testing."

Click on the Run button to execute the script. It may be easier to unit test while an ArcView document is active. Create a temporary button for that document and associate your script to it. Then unit test the script by clicking on the temporary button.

Once you are satisfied that the script is running correctly, you can associate it with a control object as explained in Chapter 4, or call it from another script as discussed in Chapter 3.

Fixing Run-Time Errors

Correcting run-time errors is more difficult. Some run-time errors will cause the script to crash during execution, while others produce erroneous results.

Use the Run button to execute your script. If the script stops before completion, a run-time error caused the crash. The following code segment shows how an invalid parameter can cause a programming crash.

```
' Code segment causing run-time crash.
thisProject=av.GetProject
aView=thisProject.FindDoc("MARYLAND")
if(nil=aView) then
if( MsgBox.YesNo("Create view?","",2) ) then
' Valid values for the third parameter of
' MsgBox.YesNo are True and False. The compiler
' does not catch this error.
newView=View.Make
newView.SetName("MARYLAND")
thisProject.Add(newView)
else
exit
end
```

Undefined classes or invalid requests can also halt the execution of a program. ArcView does not allow the creation of new classes or new services for existing classes. If you are not sure whether the class you have identified is defined, you should check the class hierarchy in the ArcView help system to verify the class and valid requests for the class.

These types of error—invalid parameters, undefined classes, and invalid requests—can be located and resolved by using the Step button to step through the code.

Run-time errors that cause unexpected results are caused by program logic errors. The best approach for correcting this type of problem is a walk-through. Using the Step button, walk through the code. At critical points, use the Examine Variables button to review variable values.

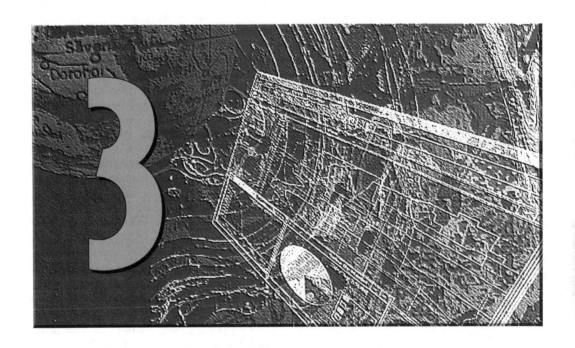

Avenue
Programming
Language

To become fluent in any language you must first master basic elements such as syntax and grammar. This chapter presents Avenue's basic programming elements, which include variables and control statements. If you have studied other programming languages, Avenue language elements will be familiar. Even if you are not a programmer, Avenue elements are easy to learn.

Programming Elements

Most programming languages have common elements. If you are familiar with other programming languages, you can learn Avenue by recognizing equivalent elements, such as the conditional If statement or the repeating While loop.

Referencing Objects with Variables

Your application can guide ArcView to execute many kinds of tasks, such as displaying a map or opening the maps table or a chart. To accomplish these and other tasks, the program needs information such as a view name or type of chart. These types of information are stored in *variables*.

Variables in Avenue are references to objects. A variable name is a sequence of letters and digits starting with a letter. Although Avenue is not case sensitive, a common practice is to begin a variable name with a lower case letter and capitalize subsequent words. For example, *primeNumber, activeDocument,* and *vehicle* are all valid variable names. There is no limit to the number of characters in a variable name.

You should verify that variable names do not conflict with class names or other reserved words. (See

Appendix B for a list of Avenue's reserved words.) The compiler produces an error message when a reserved word is used as a variable.

Next, a variable's scope is determined by its availability. For example, the scope of a local variable is limited to its script. The local variable exists only in its script; outside the script its value is not known. On the other hand, the scope of a global variable is the application. Once a global variable is created, its value is known to all scripts. Global variables are used to share objects between scripts. Any variable beginning with an underscore (_) becomes a global variable (e.g., _symbolList and _inputFileName). When a script ends, the memory area allocated to local variables is returned to ArcView. Only the referencing variable is removed from memory, and not the object itself. In contrast, the memory area allocated to global variables must be cleared and returned to ArcView by issuing the following statement.

```
av.ClearGlobals
```

If you try to read the value of a global variable after issuing the ClearGlobals request, you get an error message stating that the global variable no longer exists. ArcView remembers the global variable names, but not their values, between sessions. When you open a project, existing global variables in that project are set to *nil*. (More detail on nil below.) You can initialize global variables in a project start-up script.

✓ **TIP:** *You cannot compile a script with a global variable that is not initialized.*

Avenue does not require declaration of a variable before it is used. When you assign an object to a variable, Avenue initializes the variable. To assign an object to a variable, use one of the following formats.

```
variableName = object
variableName = expression
```

Avenue evaluates expressions and returns an object. If a variable appears on the right side of the equal sign before it is initialized, the compiler produces an error message. The following code lines show examples of variable assignment.

```
roadClass = "HEAVY"
unitReplacementCost = 5601.45
analysisBaseYear = Date.Make("1985","yyyy")
```

✓ **TIP:** *Avoid meaningless variable names such as* a1 *or* bb. *Program maintenance is easier when you use meaningful names such as* landUseList *or* roadwayTheme.

Using Object Tags

An alternative to using global variables is the object tag. You can attach an object to another that has a scope beyond the running script. Only certain objects accept tags. For example, you can tag an alias name to a theme by tagging a string object. The following code segment shows how to tag and retrieve an object.

```
aTheme.SetObjectTag ("Alias Name")
aliasName = aTheme.GetObjectTag
```

A detailed discussion of how to use tags appears in this chapter under the section "Using Bitmaps."

Variable Data Types

Avenue models data types through objects. Date, String, and Number are common data types. Others include Boolean, EnumerationElt, Nil, and Pattern. The following diagram shows the part of an object diagram that relates to data types.

Avenue data types

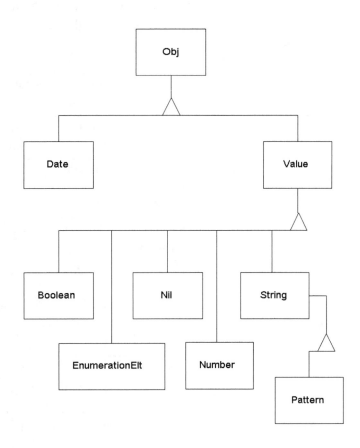

↝ **NOTE:** *Avenue's class hierarchy starts with the Obj class. This is an abstract class that provides basic requests such as GetName.*

Data types are described in the following sections.

Using Dates

Date represents a point in time and includes information on both date and time. You should become familiar with date formats before using them. A date format consists of date components and separators. A separator can be any printable character except the characters used by the date components listed in the following table.

y	The year without century from 1 to 99.
yy	The year without century from 01 to 99.
yyy	The same as yyyy.
yyyy	The year with century.
M	The month as an integer from 1 to 12.
MM	The month as a two-digit integer from 01 to 12.
MMM	The short form of the month's name.
MMMM	The month's name.
d	The day of the month as an integer from 1 to 31.
dd	The day of the month as a two-digit integer from 01 to 31.
ddd	The short form of the day's name.
dddd	The day's name.
m	The minute as two-digit integer from 00 to 59.
s	The second as two-digit integer from 00 to 59.
h	The hour from 1 to 12.

hh	The hour as two-digit integer from 01 to 12.
hhh	The hour from 0 to 23.
hhhh	The hour as two-digit integer from 00 to 23.
AMPM	The AM or PM string.
TZ	The time zone string.

➼ **NOTE:** *Date components in date format are case sensitive.*

You can use the SetFormat request to establish the format. When you send this request to a date object, you only affect the format of that date object. You can set the default format for all dates by sending the SetDef-Format to the date class. An example of how to use dates appears below.

```
yesterday = Date.Now - (1.AsDays)
yesterday.SetFormat ("dddd MMMM d, yyyy")
MsgBox.Info ("Yesterday was"++yesterday.AsString,"")
```

03-01.AVE

The preceding script displays the yesterday object in a format similar to "Thursday August 8, 1996." The Now request creates a date object with the system date and time. You can create date objects with specified values as shown in the following statements.

```
ID4 = Date.Make ("7/4/97","M/d/yy")
ID4 = Date.MakeFromNumber (19970704)
```

When you use the Make or MakeFromNumber requests, the format for the new date object is also set. In the case of MakeFromNumber, the format is set to *yyyy:MM:dd*. The parameter for the MakeFromNumber request must be a number in the *yyyyMMdd* format.

You can compare dates or apply mathematical operators. When you perform arithmetic operations on
dates, you may create a duration object. Objects of the
duration class model a span of time. A duration may
be anchored with specific starting and ending dates,
or may be floated from some arbitrary point in time.
Durations are explained in more detail later in this
chapter.

Using Booleans

Objects of the Boolean class model True and False
concepts. These objects are created by either a direct
assignment using true or false keywords, or by evaluating a Boolean expression as shown in the following
code segment.

```
aTrueObj = true
aFalseObj = false
iAmYounger = (myBirthDay > yourBirthDay)
MsgBox.Info (iAmYounger.AsString,"")
```

The AsString request results in a True, False, or Boolean null string. The null is created when you read a
Boolean-type table field and no data exists in that
field. Use the IsNull request to determine whether a
Boolean object is null.

Using Enumerations

Many of the requests in the ArcView class hierarchy
have parameters, or return an object that specifies a
value from a predefined set. For example, the Get-
Type request for a field object returns the data type of
a field. In the same manner, when you create a field
using the Make request, you must supply the field's

data type from the list of available data types. A character field is created in the following example.

```
newField = Field.Make ("LAST_NAME", #FIELD_CHAR, 40, 0)
```

The #FIELD_CHAR is the enumeration that specifies character field type. All enumerations start with the pound (#) sign. Although they are not case sensitive, you should adopt the standard of using upper case letters. To obtain a list of enumeration objects within a set, use the ReturnElements class-request as shown in the following example.

```
MsgBox.ListAsString
(EnumerationElt.ReturnElements("FieldEnum"),"","")
```

Using Nil

It is possible that attempting to assign an object to a variable could fail due to an error or non-existence of the object. In such cases Avenue assigns a nil object to the variable. This object—the sole instance of class nil—is represented by the keyword *nil.* You can use this keyword to access the nil object as shown in the following example.

```
myView = av.GetProject.FindDoc ("View1")
if (myView = nil) then
   MsgBox.Error ("View1 does not exists","")
end
```

✓ *TIP: Checking for nil objects gives you the ability to handle errors gracefully.*

Nil is different from null. Some objects such as dates or shapes may be null, which means they have no value. Test for null value by using the IsNull request.

Using Numbers

ArcView models numerical values through the instance of the number class. Although this class does not distinguish between integer and real values, you can use the format and certain requests to implement a number as an integer. Format can be applied to a single number object or the default for all numbers. You can use the SetDefFormat or SetFormat to establish a new numerical format. Both requests require a format string as a parameter. The format string consists of one or more lowercase *d* letters, and an optional decimal point character. The following code segment presents examples of setting this format.

```
aNum = 20.9
aNum.SetFormat ("d.d")     'results in 20.9
aNum.SetFormat ("d")       'results in 21
aNum.SetFormat ("dddd.dddd")   'results in 0020.9000
newNum = aNum. Truncate    'results in 20
```

CODE   03-02.AVE

Setting the format affects how AsString and AsNumber requests perform.

✓ **TIP:** *Avenue evaluates chained mathematical requests from left to right unless parentheses are used to establish the order of precedence.*

```
theAnswer = 2 + 8 / 4    'results in 2.5
theAnswer = 2 + ( 8 / 4 )   'results in 4
```

Using Strings

Objects of the string class represent character data types. Strings are of fixed length and may contain printable and non-printable characters. You can send the AsChar request to a number in order to convert the number to its corresponding ASCII character. Avenue provides keywords for the following three non-printable characters.

- ❐ NL (new line)

- ❐ TAB (tab)

- ❐ CR (carriage return)

You can embed these characters in a string using one of the concatenation operators, + or ++. The ++ operator inserts a space between the two concatenated strings. An example of concatenation appears below.

```
MsgBox.Info ("First line"+NL+"Second line","")
```

A string object is often used to create objects of other classes as shown in the following code segment.

```
aNum = "7".AsNumber

Date.SetDefFormat ("M/d/yyyy")
aDate = "8/9/1996".AsDate

aFilename = "c:/tmp/log.txt".AsFilename
```

ArcView provides several requests that allow you to manipulate or examine a string. The following code segment demonstrates such requests.

```
upperCase = yourSelection.UCase
noBlanks = myString.Trim
myQuote = simpleString.Quote
foundPoint = yourAnswer.Contains("point")
```

Using Patterns

Patterns are specialized strings that facilitate string comparison. You can create a pattern by using the Make request. The asterisk (*) and question mark (?) symbols have special meaning in patterns. Zero or more characters can replace the asterisk, while the question mark represents a single character. The following example demonstrates the use of patterns.

```
aString = "Ali Baba"
aPattern = Pattern.Make ("Ali*")
matched = (aString = aPattern)    'results in true
```

In the following example the string and pattern do not match, because the question mark requires one and only one character to be at that position.

```
aString = "Ali Baba"
aPattern = "Ali?".AsPattern
matched = (aString = aPattern)    'results in false
```

If you need the asterisk or question mark character in the pattern, but not as a special character, prefix it with a backslash.

Statements

Avenue scripts are developed by writing Avenue statements. Among statement types, *assignment* is the most common. The assignment statement consists of an equal sign with a variable on its left and an expression or object to its right. The result of the expression or object on the right is then assigned to the variable on the left. Other types of statements are explained in subsequent sections.

Controlling Program Flow

Controlling the logic flow in any programming language is a basic operation. Such controls range from executing a set of statements if a certain condition prevails, to executing the same set more than once as long as a particular condition exists. In all controls, the flow direction is based on a Boolean condition. The Boolean class includes True and False instances. The True and False objects are used for testing the condition of a control statement.

Boolean expressions are any sequence of requests resulting in True or False objects, and are used by the control statements to direct program flow. Compound expressions are created by using AND and OR requests. Examples of Boolean expressions appear below.

```
( themesList.Count > 0 )
' The above expression evaluates to True if
' the themesList list has any members.
'
( aStringList.FindByValue("STATE")>>=0 AND
(aStringList.FindByValue("COUNTY")>>=0) )
' If both STATE and COUNTY string objects are
' members of aStringList, then it evaluates to True.
'
' The following Boolean expressions are True.
(True)
(NOT False)
'
' The following Boolean expression is False.
("ARC" = "VIEW")
```

Control statements are divided into conditional and loop structures.

Conditional Statement

An If statement is used to conditionally execute a series of statements. The condition is a Boolean expression resulting in True or False. For instance, you may want to display a theme's legend only if the theme is displayed.

The conditional statement is enclosed inside parentheses and may contain more than one Boolean expression. Use parentheses to establish precedence in evaluating compounded Boolean expressions. Otherwise, compound Boolean expressions are evaluated from left to right.

The expressions on both sides of an AND request must be true to obtain a true result. One or both expressions on either side of an OR request must be true to get a true result. The NOT operator reverses the expression result. The following code segment shows examples of If statements.

CODE   03-03.AVE

```
' This script displays State and County themes
' in an open view named USA.
thisProject = av.GetProject
thisView = thisProject.FindDoc("USA")
themesList = thisView.GetThemes
stateThemeNumber = -1
countyThemeNumber = -1
anIndex = 0
for each aTheme in themesList
if (aTheme.GetName = "State") then
    stateThemeNumber = anIndex
    elseif (aTheme.GetName = "County") then
    countyThemeNumber = anIndex
    end
    anIndex = anIndex + 1
```

```
end
stateTheme = themesList.Get (stateThemeNumber)
countyTheme = themesList.Get (countyThemeNumber)
if (Not (countyTheme.IsVisible)) then
    countyTheme.SetVisible (True)
end
if (Not (stateTheme.IsVisible)) then
    stateTheme.SetVisible (True)
end
```

An If structure starts with the statement

```
if (condition) then
```

and it can have optional statements

```
elseif (conditional) then
```

or

```
else
```

and must end with the statement

```
end
```

If structures can be nested. A nested If structure is a conditional block inside another If block. If the condition is true, Avenue executes the program lines following the If statement until it reaches End, Else, or Elseif statements. All If structures end with an End statement. If the condition is false, Avenue moves to the first Else, Elseif or End statement. In an If structure, you can include many Elseif lines, but only one Else statement. Always use the Else statement after all Elseif statements. When Elseif and Else statements are part of an If structure, the Else block is executed only if none of the Elseif conditions are true.

✓ **TIP:** *Whenever possible, avoid complex nested and compound If statements. They can be the source of run-time logic errors.*

Loop Structure

Loop structures execute a set of code lines more than once. For example, you may write one set of code to display a theme, and then iterate the statements to display all themes. Avenue offers While and For Each loops.

The structure of While loops is presented below.

```
While (condition)
' Code lines
End
```

If the condition is true, Avenue executes the statements inside the loop structure; otherwise, the program flows to the End statement. When Avenue encounters the loop, it first evaluates the condition. If True, it executes the loop structure once, and returns to the top of the loop. Avenue evaluates the condition before executing every loop.

Loops can be nested. A nested loop is a loop inside another loop. Although Avenue has no limits on the number of nested loops, limits are imposed by computer memory size.

The following code segment shows an example of the While loop structure.

```
while (MsgBox.YesNo("Re-run the script?","", False))
   av.Run("myScript",nil)
end
```

The For Each loop structure works with a list or range of values. The structure of For Each loops appears below.

```
For Each name in list
' Code lines
End
```

A range can be defined by the infix operator; for instance, 0..4 provides a range from 0 to 4. ArcView models a range of numbers as instances of class interval. Intervals are discussed later in this chapter. Avenue executes the code lines inside the loop structure for every value in the list. To skip list values, you can add the word *by* to the loop statement. The For Each loop can also be nested. The following code segment shows an example of the For Each loop structure.

```
' Assuming that for every bridge class in the bridgeClass
' list, the unit maintenance cost, average area, and
' life span are available in various dictionaries,
' This loop creates a dictionary for the annual
' maintenance cost.
bridgeMaintenanceCost = Dictionary.Make (2)
for each bridgeType in bridgeClass
    cost = bridgeUnitCost.get(bridgeType) *
    (bridgeAvgArea.get(bridgeType)) /
    (bridgeLifeSpan.get(bridgeType))
    bridgeMaintenanceCost.Add(bridgeType,cost)
end
```

✓ **TIP:** *When coding a loop structure, always verify how the loop ends. An endless loop will hang your application.*

The program flow inside any type of loop structure can be altered by using Break and Continue statements.

A Break statement causes an immediate exit from the loop. If the Break statement is inside a nested loop, it affects only the innermost loop containing the statement. Break is often used when there is more than one reason to end the loop.

A Continue statement causes Avenue to skip the rest of the loop and go to the next iteration of the loop. Use the Continue statement to substitute for lengthy If statements in a loop structure. The code segment below shows an example of a Continue statement in a loop structure.

```
' Let's copy up to five zone codes from a
' master list to a new list excluding the R1 zone.
newZone = List.Make
masterCount = masterZoneList.Count
for each index in 0..4
   if (index >= masterCount) then
' The loop has passed the end of the master list.
   Exit
   end
   aZone = masterZoneList.Get(index)
   if (aZone = "R1") then
'   An R1 zone detected: skip the rest of the loop.
   Continue
   end
   newZone.Add(aZone)
end
```

Documenting with Comment Lines

Documenting your program can save time in the future when you need to maintain the application code. The Avenue character for comments is a single quote ('). This symbol in program code appears in previous sections. Wherever this token appears, everything to the end of the line is read as a comment. The exception is when a single quote appears inside a string. The following code segment shows examples of a comment token placed at the beginning and the middle of a line, and an instance of where the token does not precede a comment.

```
' This comment starts at the beginning of a line.
theView = theProject.GetActiveDoc ' Comments go here.
MsgBox.Info("I miss the 70's","")
' The single quote in the preceding statement does
' not create a comment line.
```

✓ **TIP:** *Never procrastinate about documenting your code. Chances are that you will not do it later. Always start your program with comment lines containing your name, date, program name, version number, and purpose.*

Using Collections

In object oriented languages, collections are objects that contain other objects. They are useful in organizing related information. For instance, it is easier to keep the names of all your themes in a list called *allMyThemes* than scattered among many different variables.

Using Lists

Avenue provides several classes of collections. Lists and dictionaries are the most common types. Lists are series of objects created through direct assignment or by request. They can contain objects of different classes. The following code segment shows how lists are created.

03-04.AVE

```
roadClass = {"LIGHT", "MEDIUM", "HEAVY"}
bridgeClass = roadClass
bridgeClass.Add ("DAMAGED")
' The bridgeClass list is composed of LIGHT,
' MEDIUM, HEAVY and DAMAGED string objects.
roadAnnualCost = List.Make
' The Add request appends an object to the end of a list.
roadAnnualCost.Add (25000)
' The Insert request places an object on the top of a list.
roadAnnualCost.Insert (12000)
lightRoadAnnualCost = roadAnnualCost.Get (0)
' The variable lightRoadAnnualCost references
' the number object 12000.
```

In Avenue, requests that can return more than one object use a list to organize the returned objects. In the following example, the GetThemes request returns a list containing all themes of a view.

```
myView = av.GetProject.FindDoc ("View1")
allMyThemes = myView.GetThemes
firstTheme = allMyThemes.Get(0)
```

Using Dictionaries

Dictionaries are an efficient method to store and retrieve objects. In concept, they are similar to a two-column table with identifier and value column headings. Any type of object can be stored by a key, and the key may be from any class. Dictionaries accept objects and keys of different classes. The following code segment shows an example of how to use dictionaries.

 03-05.AVE

```
roadTravelCost = Dictionary.Make (2)
' The (2) parameter refers to the size of
' hash tables and not number of objects in
' the dictionary. The number of objects should
' not exceed hash size times 20.
roadTravelCost.Add ("HEAVY",35.48)
roadTravelCost.Add ("MEDIUM",35.00)
roadTravelCost.Add ("LIGHT",34.28)
roadTravelCost.Add ("GRAVEL",35.50)
gravelRoadCost = roadTravelCost.Get("GRAVEL")
' The variable gravelRoadCost references
' the number object 35.50.
```

✓ **TIP:** *Dictionaries are efficient for random access, while lists are efficient for sequential access.*

Object Model for Collections

The following diagram shows the object model for collection classes.

Collections object model.

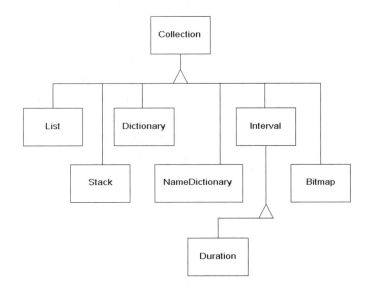

Collection is an abstract class that defines the common attributes and operations among its subclasses. Collections can be divided into ordered and unordered. Bitmap, Interval, Duration, List, and Stack are ordered while Dictionary and NameDictionary are unordered. The difference is that the elements of ordered collections can be accessed in a specific sequential order.

ArcView does not impose any size limit on the objects of collection sub-classes. However, when you create a bitmap you need to establish its size.

Using Stacks

Stacks are similar to lists, but stack elements are not accessed in the same way. Stacks implement a last-in-first-out access method in which the stack elements are retrieved and removed in the reverse order they were added. They are useful in implementing undo logic. For instance, you can keep track of selected features as they change and allow undoing the selection. See the section on "Using Bitmaps" in this chapter for an example of using stacks.

Using Name Dictionaries

Objects of NameDictionary are similar to dictionaries, except that the element's name becomes the key. A name dictionary is more efficient than a standard dictionary. Make sure that each object has a name before adding it to the name dictionary. Since the object's name is the key, no two elements can have the same name.

✓ *TIP: You cannot access an element if you change its name. You must remove the object from the name dictionary, change its name, and then add it again.*

The following code segment shows how you can create a name dictionary.

```
myNDict= NameDictionary.Make (2)
      ' Add request rejects the object if another
      ' element with the same name exists.
myNDict.Add (locatorView)
      ' Set request overwrites any existing
      ' element with the same name.
myNDict.Set (indexView)
```

Using Intervals and Durations

Objects of interval class model a range of numbers with a defined set of subdivisions. Intervals are often used to construct a For Each loop. If the loop requires iterating through a range of numbers in equal increments, an interval is more efficient than a list of numbers. An interval object maintains its starting, ending, and incrementing values, and calculates the value of each subdivision. The starting and ending values are also known as the lower and upper bounds, but the lower bound need not be smaller than the upper bound. The incrementing value is set to 1 but can be changed. The following code segment is an example of intervals.

03-06.AVE

```
' You can create an interval
' with the .. operator.
oddInterval = 1..9 by 2
for each oddDigit in oddInterval
   MsgBox.Info ("Odd digit:"++oddDigit.AsString,"")
end
```

✓ **TIP:** *The default increment value is -1 when the starting value is greater than the ending value.*

Durations are specialized intervals to model a range of dates. Durations are either anchored with specific start and end dates or floated to represent a span of time. You can create anchored durations with the double period (..) operator. The loop variable for an anchored duration is an instance of class date. The following script is an example of anchored duration.

03-07.AVE

```
' This script creates and displays
' a list of Mondays in 1996.
mondayInterval = (19960101.AsDate)..
(19961230.AsDate) by 7.AsDays
mondayList = {}
for each monday in mondayInterval
   mondayList.Add (monday)
end
MsgBox.ListAsString (mondayList,"Mondays in 1996","")
```

You can create a floating duration with the minus operator. The loop variable for a floating duration is a number.

➛ **NOTE:** *The default increment for a duration is one second.*

Using Bitmaps

Bitmaps are collections of Boolean values. They are similar to lists except that each element is a Boolean object. Next, objects of a Bitmap class have a pre-defined size. Although you can create a bitmap with a Make request, you typically work with the existing bitmaps. ArcView uses bitmaps to identify spatial or attribute records that are selected or defined as part of a query.

The following two scripts provide the ability to undo feature selection in a view document. Chapter 4, "Customizing the Interface," discusses adding new buttons and assigning Avenue scripts to the buttons. The scripts listed here are assigned to the update and click properties for undoing selections. The script assigned to the update property is executed by ArcView every time a feature is selected. The script assigned to the click property is executed when the button is selected.

03-08.AVE

```
' View.UndoSelUpdate
' This script is assigned to the update
' property of selection undo button.
theView = av.GetActiveDoc
themeList = theView.GetThemes
'

' Determine if the selection has changed.
addIt = False
for each aTheme in themeList
    '

    ' Disregard image themes.
    if (aTheme.Is(FTheme)) then
        '

        ' Each selection bitmap is stored in a stack
        ' tagged to each theme. Add the stack
        ' if it does not exist.
        theStack = aTheme.GetObjectTag
        if (theStack = nil) then
            theStack = Stack.Make
            aTheme.SetObjectTag (theStack)
        end
        '

        ' Compare the current and last selections.
        selectionBitmap = aTheme.GetFtab.GetSelection
        if (theStack.IsEmpty) then
            if (selectionBitmap.Count = 0) then
                ' Nothing selected.
                continue
            else
                addIt = True
            end
        else
            lastBitmap = theStack.Top
            if (lastBitmap = selectionBitmap) then
                ' Selection has not changed.
                continue
```

```
                            else
                                addIt = True
                            end
                    end
                end
        end
        '
        ' Add it to the stack of each theme if
        ' the selection has changed.
        if (addIt) then
            for each aTheme in themeList
                if (aTheme.Is(FTheme)) then
                    theStack = aTheme.GetObjectTag
                    SelectionBitmap = aTheme.GetFtab.GetSelection
                        ' Add a copy of the bitmap to the stack.
                    theStack.push (selectionBitmap.Clone)
                end
            end
        end
```

CODE · 03-09.AVE

```
' View.UndoSel
' This script is assigned to the click
' property of selection undo button.
theView = av.GetActiveDoc
themeList = theView.GetThemes
'

' Set the selection to the bitmap
' stored in a stack. The stack is
' tagged to each theme.
for each aTheme in themeList
        '
        ' Disregard image themes.
        if (aTheme.Is(FTheme)) then
            theStack = aTheme.GetObjectTag
            selectionBitmap = aTheme.GetFtab.GetSelection
            if (theStack = nil) then
```

```
                                      '
                                      ' New themes may not have object tags,
                                      ' which means no feature was selected
                                      ' in the last change to the selection.
                                      theStack = Stack.Make
                                      aTheme.SetObjectTag (theStack)
                               end
                               if (theStack.Depth < 2) then
                                      selectionBitmap.ClearAll
                                      theStack.Empty
                               else
                                      ' The top element of the stack matches
                                      ' the displayed selection. Discard
                                      ' the top bitmap, then use the next one
                                      ' without removing it from the stack.
                                      theStack.Pop
                                      lastBitmap = theStack.Top
                                      selectionBitmap.Copy(lastBitmap)
                               end
                               aTheme.GetFtab.UpdateSelection
                        end
                  end
```

Interaction Between Programs

If your application is small and narrow in scope, a single Avenue script may be sufficient. Most applications, however, require interaction among several Avenue scripts, and skillful developers design applications in modular form. In a modular design, application functions are broken down into separate Avenue scripts. Maintainability and reusability are among the major benefits of modular design.

However, modular design introduces a new challenge to the application developer: how will the program modules interact? The two principal types of interaction are calling other programs and sharing information.

In a typical modular design, a master script controls the entire application and calls upon other scripts to carry out specific functions. The master script is often called the *main program*, and the scripts it calls are known as *subroutines*. Subroutines perform specific tasks using information supplied by the main program. The next two sections explain how the main program calls its subroutines and shares information with them.

Calling Other Programs

An Avenue script can start another script by using the Run request. The format of such a request follows.

```
av.Run (scriptName, selfObject)
```

Avenue searches first for the script identified by *scriptName* in the current ArcView project file. If the script is not found in the current project, Avenue searches the user and system default projects. Avenue displays an error message if the script is not found. Once the subroutine has completed its task, Avenue returns to the next statement in the main program. The following code segment is an example of the Run request.

```
' Present a list of possible projections.
prjList = "Lambert Albers Azimuthal."AsList
whichPrj = MsgBox.ChoiceAsString
(prjList,"Select a projection: ","")
' Stop the program if user clicks on
' the cancel button.
```

```
if (nil = whichPrj) then
    exit
end
whichPrj = whichPrj.UCase
' Call appropriate subroutine to
' perform the projection.
if (whichPrj = "LAMBERT") then
    av.Run ("LAMBERT_PRJ","")
elseif (whichPrj = "ALBERS") then
    av.Run ("ALBERS_PRJ","")
else
    av.Run ("AZIMUTHAL_PRJ","")
end
```

If you plan to use a script from the user or system defaults, you should first verify that the script exists. If the script you need is not available, you can prevent your application from crashing by using the FindScript request to carry out a search. An example appears below.

03-10.AVE

```
aScript = av.FindScript ("ALBERS_PRJ"
if (nil = aScript) then
    MsgBox.Error
    ("Unable to find the script.","")
    exit
else
    av.Run ("ALBERS_PRJ","")
end
```

If the script is located, FindScript returns the script name. FindScript returns nil if the script is not located.

✓ **TIP:** *In a "top-down" approach to application development, you begin by writing the main program to call subroutines that perform various functions. The main program could be your* *application's main menu. Under this approach, each function is called based on the user's selection from the main menu.*

Passing Objects Between Programs

A subroutine often requires information from the main program to complete its task. For instance, the LAMBERT_PRJ subroutine mentioned in the previous section requires parameters such as central meridian, reference latitude, and standard parallels for a Lambert Conformal Conic projection. The main program will also need to know if the projection was carried out successfully when LAMBERT_PRJ ends.

The selfObject in the Run request (see the previous section, "Calling Other Programs") provides the means to share information between the main and subroutine programs. By using the keyword "Self," a subroutine references its selfObject. Storing shared data in a dictionary and using the dictionary as the selfObject is a useful practice. The following code segments illustrate this method. The first block appears in the main program, and the second block appears in the subroutine.

```
' Store predefined projection parameters
' in the main routine.
sharedDictionary = Dictionary.Make (2)
sharedDictionary.Add ("CM",cenMeridian)
sharedDictionary.Add ("RL",refLatitude)
sharedDictionary.Add ("P1",firstParallel)
sharedDictionary.Add ("P2",secondParallel)
sharedDictionary.Add ("SUBJECT",theView)
sharedDictionary.Add ("EXTENT",theExtent)
```

```
sharedDictionary.Add ("SUCCESS",False)
' Run the subroutine that performs the projection.
av.Run("LAMBERT_PRJ",sharedDictionary)
' Check to see if projection was completed.
if (Not (sharedDictionary.Get("SUCCESS"))) then
    MsgBox.Warning
      ("Unable to perform the projection","")
end
```

```
' Create and set the projection in the subroutine.
' Refer to sharedDictionary as SELF.
thisLambert = Lambert.Make(Self.Get("EXTENT"))
thisLambert.SetCentralMeridian (Self.Get("CM"))
thisLambert.SetReferenceLatitude (Self.Get("RL"))
thisLambert.SetLowerStandardParallel (Self.Get("P1"),
thisLambert.setUpperStandardParallel (Self.Get("P2"))
thisLambert.Recalculate
thisView = Self.Get("SUBJECT")
thisView.SetProjection (thisLambert)
Self.Remove ("SUCCESS")
Self.Add ("SUCCESS",True)
```

✓ **TIP:** *You do not have to develop all subroutines before testing the main routine. Create a skeleton with a message box to show that a particular subroutine was called. This process is similar to prototyping, or walking the user through your application without real-time task execution.*

Values can be returned to the calling script by using the Return statement. If you do not specify a Return statement, the last object created in the subroutine is returned. The following is an example of the Return statement.

```
' In the main program:
worked = av.Run ("LAMBERT_PRJ",sharedDictionary)
' In the subroutine:
return True
```

Accessing Files

Application files can be sources of input, temporary storage areas, output destinations, or all three. Avenue recognizes text and line files, and both file types store printable characters. Therefore, the number 123 is stored as the three characters, 1, 2, and 3. Generally, each character corresponds to one byte, except in some non-English versions of operating systems. Avenue reads and writes the contents of text and line files differently. Avenue accesses a text file one character at a time, while line files are accessed one line at a time. A line is a series of characters ending with a platform-dependent delimiter. (In Microsoft Windows, the delimiter is comprised of both a carriage return and newline character; in Macintosh, only a carriage return character; and in the UNIX environment, the newline character.) The length of each line is limited only by memory size. The following diagram shows the object model for the files.

*Object model
related to files.*

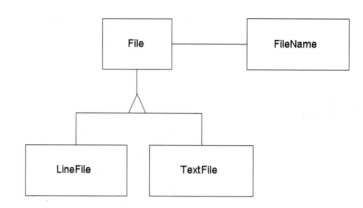

To read a file or write into it, you must first open or create the file by using a file name object. The file name object and associated tasks are described in the next section.

The File Name Object

Avenue stores file names as objects of the FileName class. A thorough understanding of this class is needed for using requests that handle file objects. FileName is an operating system-independent class that, through its instances, refers to valid file names in various operating systems. The file may or may not physically exist.

Before using any file, a FileName class object should be created. The Make request, as shown below, creates a FileName object.

```
myFilename = FileName.Make ("POP90.DAT")
```

In this example, *myFilename* becomes a file name object, and POP90.DAT is the object's value. Requests to the FileName class also deal with the file path. In the following example, a file name is created with the system's temporary directory as its path.

```
tmpDirName = FileName.GetTmpDir
myFilename = FileName.Make ("AVGDATA.TMP")
myFilename = FileName.Merge (tmpDirName, myFilename)
```

An alternative is to replace the last two lines of the above with the next two lines.

```
myFilename = FileName.Merge (tmpDirName,
"AVGDATA.TMP".AsFilename)
```

If you know the temporary path, all three lines can be replaced with the single statement below.

```
myFilename = FileName.Make("C:\Windows\temp\AVGDATA.TMP")
```

In multi-user applications, unique names for each user's temporary files must be assigned to prevent confusion. To create a unique name, you can use the MakeTmp request as illustrated below.

```
' A prefix of up to five characters and a file extension
' can be established for unique file names.
aUniqueFilename = tmpDirName.MakeTmp ("AHR","TMP")
```

The first string is the prefix, and the second string is the extension for a unique file name. A prefix or extension could identify the application that has generated the file.

Creating, Opening, and Closing a File

The Make request to the TextFile or LineFile classes creates new files or opens existing files. When a new file is created, it is also opened. The following examples illustrate opening files.

03-11.AVE

```
' Create a text file to write to and read from;
' if file exists, overwrite it.
myTextFile = TextFile.Make
(myTextFilename, #FILE_PERM_WRITE)
' Open a line file for reading.
myLineFile = LineFile.Make
(myLineFilename, #FILE_PERM_READ)
```

The first parameter for the Make request is an instance of the FileName class, which assigns the file name. The second parameter establishes the file permissions. Permissions are listed below.

❏ #FILE_PERM_READ opens an existing file for reading.

❏ #FILE_PERM_WRITE creates a new file for writing. If the file exists, it is overwritten.

❏ #FILE_PERM_MODIFY opens an existing file for reading and writing. You must use the Flush request after each write, and the SetPos request before each read.

❏ #FILE_PERM_APPEND opens an existing file for appending. If the file does not exist, a new file is created.

❏ #FILE_PERM_CLEARMODIFY opens a new file for reading and writing. If the file exists, it is overwritten. You must use the Flush request after each write, and the SetPos request before each read.

A file object created in this manner can accept requests to read from or write to. Avenue generates an error message if file access conflicts with file permissions.

Open files may be damaged if the system crashes. Therefore, you should open a file only when it is needed and close it as soon as it is no longer needed in the application. Because opening and closing files are time-consuming, you should avoid unnecessary calls.

Once a file is closed, it cannot be used again until it is reopened. To close a file, make the Close request to your file object, as shown below. Avenue closes all open file objects at the end of your application.

```
anyFile.Close
```

✓ **TIP:** *Multiple instances of a file can be opened by several file objects that reference the same physical file. Each file object must be closed once the operation is completed.*

You can delete a file or check for its existence by making Exists and Delete requests to the File class. Each request, as shown below, needs a file name as its parameter.

```
' Delete an old file named OLD_DATA.
oldFilename = FileName.Make ("OLD_DATA")
foundIt = File.Exists (oldFilename)
' Exists request returns True or False.
```

```
' True means that the file exists.
if (foundIt) then
     File.Delete (oldFilename)
else
     MsgBox.Info (oldFilename.AsString++
     "was not found","")
end
```

Reading and Writing a File

Before listing the requests that carry out reading and writing operations, a brief discussion of file pointers and file elements is in order. A file pointer points to a location inside a file. Read and write operations start at the position of the file pointer. When a file is created or opened, the file pointer is positioned at the beginning unless the #FILE_PERM_APPEND permission is used. In that case, the pointer is positioned at the end. Avenue moves the pointer by sending the SetPos request.

A file element refers to the quantity of data Avenue reads or writes per request. The element size is based on file type: one character for text files, and one line for line files.

Read and Write are basic requests that apply to all file types. The format for these requests appears below.

```
aFileObject.Read (dataBuffer, elementCount)
aFileObject.Write (dataBuffer, elementCount)
```

The *elementCount* is a numeric object referring to the number of elements to read from the file into the dataBuffer, or to write from the dataBuffer into the file. The *dataBuffer* is made up of a list of file element objects. For instance, the dataBuffer for a line file is a list of lines.

The following statements illustrate how a line file accepts requests to read and write complete lines.

```
aLine = aLineFile.ReadElt
aLineFile.WriteElt (aLine)
```

The first statement reads the next line and stores it in the *aLine* string object. The second statement writes the characters in the aLine object to the *aLineFile*.

After each read or write, Avenue positions the file pointer at the end of the data just accessed. As illustrated in the code segment below, the pointer's position can be changed by using requests that handle the pointer's position.

```
' Place the pointer at the beginning of the file.
aFile.GotoBeg
' Place the pointer at the end of the file.
aFile.GotoEnd
' Check to see if the pointer is at the end of the file.
atTheEnd = aFile.IsAtEnd
```

Two other requests allow you to determine the position of the pointer or the file size.

The GetPos request returns the sequence number of the element the pointer is pointing at. Elements in a file are numbered from zero. Consequently, if the Get-Pos request returns 1, the pointer is pointing at the beginning of the second element.

As demonstrated below, the GetSize request returns an integer which shows the number of elements in a file.

```
elementCount = myLineFile.GetSize
MsgBox.Info ("There are" ++ elementCount.AsString ++
"Lines in the file","")
```

User Dialog

The primary method of communication with the user is through the message box class called *MsgBox*. All requests are made to the class rather than to class instances. An object cannot be created from this class.

All message boxes are modal; the user must respond to the box before continuing. Message boxes are used to provide a message or accept an input. The action is based on the request.

Displaying Information, Warnings, and Errors

Message display requests are listed below.

```
MsgBox.Info (aMessage, aTitle)
MsgBox.Warning (aMessage, aTitle)
MsgBox.Error (aMessage, aTitle)
```

In the preceding statements, the *aMessage* and *aTitle* objects are strings that will display as part of the message box. An error, warning, or information icon is displayed in the box. The box will also contain an OK button for user acknowledgment. The next figure shows the warning message box displayed by executing the following statement.

```
MsgBox.Warning ("Unable to find the theme.","")
```

*A Warning
message box.*

✓ **TIP:** *You can omit the title by keying in two double
quotation marks. Do not use nil if you intend to
omit the title. If you place a space between the
double quotation marks, no title is displayed.*

Getting an Input

The following requests are used to solicit a Yes or No
answer from the user.

```
MsgBox.YesNo (aMessage, aTitle, defaultAnswer)
MsgBox.MiniYesNo (aMessage, defaultAnswer)
MsgBox.YesNoCancel (aMessage, aTitle, defaultAnswer)
```

The *aMessage* and *aTitle* parameters are string objects
that will be displayed in the message box. Each message box will contain YES and NO buttons. The Yes-
NoCancel request creates a message box which
includes a CANCEL button. The statement returns a
true or false object. If the YES button is clicked, the
statement returns True. If the NO button is clicked, the
statement returns False. A nil is returned when the
CANCEL button is pressed. The defaultAnswer is a
Boolean object that assigns a default button to the
box. A value of False selects NO as the default button,
and a value of True assigns YES as the default button.

Users can select the default button by pressing the Enter key.

The next illustration shows a message box created by the following code segment.

```
saveIt = MsgBox.YesNoCancel
("Save project before exiting? ",",true)
if (saveIt = nil) then
    exit
elseif (saveIt) then
    thisProject.Save
    thisProject.Close
elseif (Not saveIt)
    thisProject.Close
end
```

A YesNoCancel message box.

The following series of requests are used to ask for data input.

```
inputString = MsgBox.Input (aMessage,
aTitle, defaultInput)
inputString = MsgBox.Password
inputObject = MsgBox.Choice (aList, aMessage, aTitle)
inputObject = MsgBox.ChoiceAsString (aList,
aMessage, aTitle)
```

The Input request provides an input line, an OK button, and a CANCEL button in a message box. If the user presses the CANCEL button, this request returns nil; otherwise, the string entered by the user is returned. The *defaultInput* parameter establishes a default string for the input box. If you do not have a default value, enter two double quotation marks.

The Password request creates a message box similar to the Input request, except that asterisks are displayed instead of the characters keyed in by the user.

The Choice request creates a message box with a drop-down list of objects from the *aList* parameter. The GetName request is applied to each object of the list. If the object name is not meaningful to the user, use the ChoiceAsString request, which applies the AsString request to each object from the drop-down list.

✓ *TIP: Use the GetName request for objects which have a name property, such as themes. For objects lacking a name property, such as a date object, use the AsString request.*

Use the MultiInput request if you wish to accept more than one piece of data per message box.

Getting a File Name

By making requests to the FileDialog class, you can have the application ask the user for file names. The FileDialog class can accept Put and Show requests. Because FileDialog has no class instances, an object cannot be created from this class. The result is a modal window that lists directories and existing files, and allows disk navigation. The window also has an input area for the file name, an OK button, and a CANCEL button. Examples of requests to the FileDialog class are shown in the following statements.

```
newFilename = FileDialog.Put
(defaultFilename, namePattern, aTitle)
existingFilename = FileDialog.Show
(namePattern, patternLabel, aTitle)
```

The *defaultFilename* is a FileName class object which appears in the input box of the window, but it can be replaced with a different name. The *namePattern* is a string object that determines the file names listed in the window. For instance, the "*.*" string lists all files, and "*.APR" lists files with the *APR* extension. The *patternLabel* identifies the file type for the namePattern. If the namePattern is "*.DBF," then the patternLabel can be "DATABASE FILES." The *aTitle* is a string object appearing in the window title bar.

The Put request requires input of a new file name. If the file name already exists, Avenue asks the user to confirm replacement of the existing file. The only input the Show request permits is existing files. The following statement creates the window shown in the next figure.

```
testFilename = FileDialog.Show ("*.HLP",
"Help Files", "")
```

A File Name window.

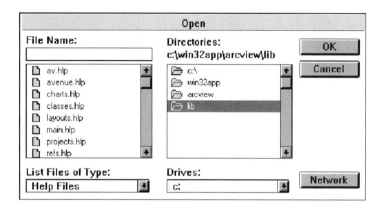

✓ **TIP:** *Changing directories through the FileDialog window also changes the current working directory.*

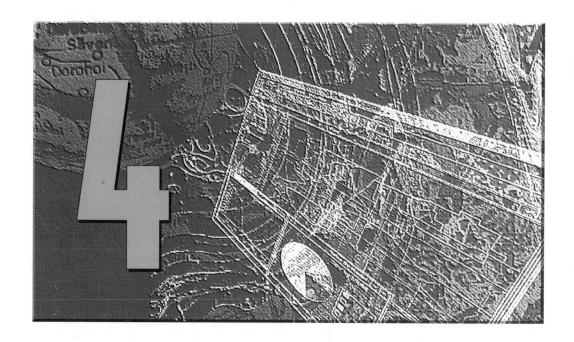

Customizing the Interface

During the process of application development, you should consider human interaction with your program, or how users will control your application through the program interface. The complexity of this interaction is directly related to the flexibility of your application. If the user is required only to start the program and provide a few parameters, the interface can be simple. However, if the user needs interactive tools to analyze and direct the application, the interface will be more complex.

Users judge a program's friendliness by its interface. If the interface does not provide feedback, or is not intuitive, users may perceive the program as unfriendly. A well-designed interface can also compensate for other application shortcomings such as slow response time.

✓ **TIP:** *You can reduce user frustration with slow response time by providing a progress status bar, or by sending messages to the status line that indicate completion of tasks.*

ArcView gives you the ability to create and customize the interface. Therefore, it is up to you to design the best interface possible for your application. You have the ability to create a new interface or modify the existing one. You can set the look and behavior of the interface. This chapter focuses on the steps in creating an interface tailored to your application. Customization can be executed interactively as part of the development process, or dynamically when your application is executed. Both methods are discussed here. However, you will find that in most instances the interface is customized interactively and stored with the project file.

What Can Be Customized

Interaction with ArcView takes place through the control bar, also known as the document GUI, or graphical user interface. The document GUI consists of three lines containing the menus, buttons, and tools, respectively. Each ArcView document has a control bar that becomes available when the document is active. The following illustration shows the document GUI for view documents.

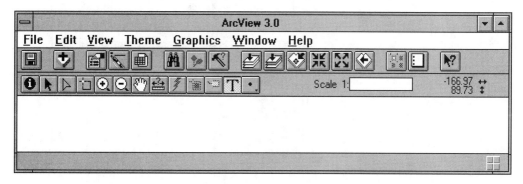

Document GUI for view documents.

The line containing menu options is known as the menu bar, while the other two lines are referred to as the button and tool bars. You can customize every interface control on the menu and button bars, along with the tools. However, the text boxes that appear on the tool bar are not customizable. ArcView has a pop-up menu displayed on most systems by clicking the right mouse button. This pop-up menu is part of the user interface for each document. You can customize the pop-up menu as well.

In addition to tailoring the user interface for ArcView documents, you can also customize the project window. The following figure shows a customized project window.

A customized project window.

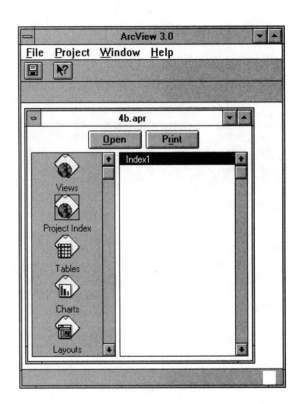

You can customize the three buttons and document icons in the project window. The buttons are known as "label buttons" because they are labeled with text instead of displaying an icon.

Menu Bar

The line of menu options is usually displayed at the top of the control bar. Additional menu items for each option are revealed by clicking the pointer on the option. Pull-down menu items offer the user an easy method of interaction. A well-designed set of menus makes an application consistent and easier to learn.

Menu names must be meaningful and indicative of their actions. You should establish and follow a set of standards for naming and grouping items. Keep in mind that users of various systems are accustomed to certain naming and grouping conventions. For instance, Microsoft Windows users generally expect File to be the first menu option, with pull-down items such as New, Open, Print, and Exit.

Button Bar

Buttons create a faster method for executing the same commands available through the menu items. Generally, button bars consist of frequently used menu items. Novice users often prefer menu options, while experienced users like buttons.

Tool Bar

The tool bar is a set of mutually exclusive tool buttons that associate a specific procedure with the cursor. Once selected, a tool remains depressed until another tool is selected. Because tool buttons are mutually exclusive, only one tool can be selected at a time. Thus, selection of a new tool deselects the current one. When a user selects a tool button, the cursor shape changes, preferably to the same icon that marks the button. A cursor shape change indicates that a tool button is active.

Pop-up Set

The pop-up set is a group of menu items related to a document. Although multiple sets of menus can be assigned to a pop-up set, only one menu set can be displayed and used at a time. A pop-up menu is best used for presenting frequently used operations in a document. You access the pop-up menu by clicking the right mouse button over an ArcView document.

Project Window

The icons listed on the left pane of the project window represent available document GUIs. ArcView has five types of document GUIs: views, tables, charts, layouts, and scripts. You can create your own type based on one of ArcView's types and add it to the project window.

Control Properties

You can customize interface controls by changing their properties. Each property is briefly described below.

❐ **Apply.** Assigns an Avenue script to a tool when the user clicks or drags the pointer on the display area of the document.

❐ **Click.** Specifies a script to execute when the user clicks on the control.

❑ *Cursor.* You can associate a cursor shape to a tool with this property. The cursor changes to the specified shape when the tool is selected.

❑ *Disabled.* A disabled control cannot be selected or used.

❑ *Help.* A short message can be displayed on the status bar when the pointer is placed on the control. The short message is placed in this property.

❑ *Help Topic.* Specifies the help topic in ArcView's help system displayed when the control is selected with the help button.

❑ *Icon.* Associates an icon with the tool or button control. The icon appears on the face of the tool or button.

❑ *Invisible.* When set to true, the control is not displayed.

❑ *Label.* Specifies the label for a menu option or item.

❑ *Shortcut.* The user selects menu items using the key combination specified in this property.

❑ *Tag.* Available to store any textual data, but it is not used by ArcView.

❑ *Update.* Specifies a script to execute when the state of the active window is changed.

Customizable properties of the document GUI appearing on the project window are briefly described below.

❏ *Action script.* Executes associated script when the user clicks on the third label button of the project window.

❏ *Action update script.* Executes associated script when the user clicks on a document GUI icon or one of its documents.

❏ *Document base name.* New documents are created using this name plus an incremental number.

❏ *Icon.* Establishes the icon shown on the left side of the project window for the document GUI.

❏ *Name.* Sets the name for the document GUI. You will use this name in Avenue to access the document GUI. If a document base name is not provided, then the name set in this property is used to create new documents.

❏ *New script.* Executes associated script when the user clicks on the first label button of the project window.

❏ *New update script.* Executes associated script when the user clicks on a document GUI icon or one of its documents.

❑ **Open script.** Executes associated script when the user clicks on the second label button of the project window.

❑ **Title.** Establishes the document GUI title appearing under the icon.

The Action script, New script, and Open script properties of the document GUI are equivalent to the Click property while Action update script, New update script, and Open update script properties are equivalent to the Update property of the buttons. The click properties are used to perform an action while update properties are used to change the state of the control. The following script shows how the third label button of script documents on the project window is enabled and gets its label. The script, named *Script.ActionUpdate*, is one of ArcView's system scripts.

```
SELF.SetEnabled (av.GetProject.GetSelectedDocs.Count>0)
SELF.SetLabel("&Run")
SELF.SetHelpTopic("Run_button")
```

The SELF keyword, when used in an update property script, represents the control object. In the preceding example, the script represents the third label button object because it is associated to the action update script property of the scripts document GUI. Therefore, the SELF.SetEnabled statement is the same as the following code segment.

```
thirdButton = av.GetProject.GetButtons.Get(2)
thirdButton.SetEnabled(av.GetProject.GetSelectedDocs.Count>0)
```

The ampersand in *&Run* sets the letter *R* as the access key for this button. Assigning access keys is discussed in more detail later in this chapter.

Event Programming

The term "event" refers to the occurrence of an action. For instance, when the user makes a theme active, an event has occurred inside the view document. Task initiation based on an event is called "event programming." User interface controls have Click, Update, and Apply event properties. An Avenue script can be associated to these properties, and the scripts are executed when a click, update, or apply event takes place.

A click event occurs when the user clicks on a control. An apply event occurs when the user clicks or drags the mouse in the display area while a tool button is active. Update events occur when the state of a document changes. For example, displaying a theme or selecting a feature triggers an update event.

Update event programming is useful in a variety of situations, although it is used mainly for enabling or disabling user interface items. For instance, you might create a tool button that computes bridge replacement cost in a road coverage. If the theme for the road coverage is not present, the button should be invisible. If the theme is present but not visible, the tool button should be disabled. The tool button should be available for use only when the road coverage is displayed.

An Avenue script for update event programming is associated with the update property of a user interface item. The script consists of two parts: testing for an event and performing a task. ArcView automatically executes the script every time an update event occurs. Subsequent sections describe how to test for different events.

Checking for a Condition

Requests with an *Is* prefix are used in checking for a condition. For example, you can use the IsActive request to determine whether a theme is active. Several of Avenue's numerous requests starting with Is are listed below.

```
aTheme.IsActive
aChart.IsChartScatter
Coverage.IsINFO (myCoverage)
aControl.IsEnabled
aString.IsNumber
aTool.IsSelected
aTheme.IsVisible
```

The above Is requests return a true or false Boolean object. The following code segment shows how a tool button is enabled when a theme is visible.

```
' This script is associated with the
' update property of the tool button.
if (theTheme.IsVisible) then
Self.SetEnabled (True)
else
Self.SetEnabled (False)
end
```

In an update script, the keyword *Self* refers to its control. Therefore, applying the setEnabled request to Self is similar to applying the request to the interface control object.

Checking for Active Objects

Requests with a *GetActive* prefix are used to access active objects. An object is made active or inactive by clicking the mouse on it. Active objects are generally displayed differently. For instance, an active theme is raised while an active field is lowered. These requests return the active object(s), or nil if there are none. A short list of GetActive requests follows.

```
av.GetActiveDoc
aTable.GetActiveField
aView.getActiveThemes
```

The following code segment shows how selecting a numeric field enables a customized menu item.

```
' This script is associated with a menu item
' for statistical analysis on the selected
' numeric field.
aField = theTable.GetActiveField
if (nil = aField) then
Self.SetEnabled (False)
elseif ( aField.IsTypeNumber ) then
Self.SetEnabled (True)
else
Self.SetEnabled (False)
end
```

Object Model for Control Bar

If you plan to change the user interface dynamically, you need to understand the object model for the user interface. The starting point for this model is the DocGUI class as shown in the following diagram.

Object model for control bar.

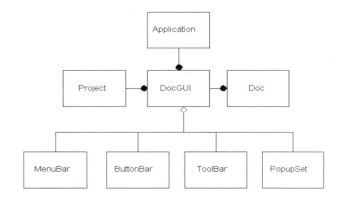

A DocGUI class models the controls used in accessing ArcView documents. Each DocGUI object is comprised of MenuBar, ButtonBar, ToolBar, and PopupSet objects.

The scrolling list on the left side of the project window shows the available DocGUIs. You can create a DocGUI by either cloning an existing one or making a new one, as shown in the following code segment.

```
clonedDocGUI = av.FindGUI("View").Clone
newDocGUI = DocGUI.Make("View")
```

You need to make an explicit request to show the cloned or new DocGUI on the project window. The following code segment shows how to add a DocGUI to the project window.

```
av.GetProject.AddGUI (newDocGUI)
```

When you create a DocGUI and add it to the project list, you should set several of its properties as shown below.

❏ Assign a unique name so it can be found.

```
newDocGUI.SetName ("IndexGUI")
```

❒ Assign a title for its icon on the project window.

```
newDocGUI.SetTitle ("Project Index")
```

❒ Assign a view icon.

```
newDocGUI.SetIcon(av.FindGUI("View").GetIcon)
```

❒ Assign the scripts that are executed when one of the buttons on the project window is selected.

```
newDocGUI.SetNewScript("Doc.New")
newDocGUI.SetOpenScript("Doc.Open")
newDocGUI.SetActionScript ("Doc.Action")
```

❒ Each button also has an update property that needs a script.

```
newDocGUI.SetNewUpdateScript ("Doc.NewUpdate")
newDocGUI.SetOpenUpdateScript ("Doc.OpenUpdate")
newDocGUI.SetActionUpdateScript ("Doc.ActionUpdate")
```

❒ Make the new document GUI visible on the project window.

```
newDocGUI.SetVisible (true)
```

If you clone the GUI you can avoid using most of the preceding code.

Creating a Menu System

When creating a menu system, you will be working extensively with the Customize dialog box. To access the Customize dialog box, activate the Project window by clicking the pointer on its title bar. Select the Project menu option from the control bar to display a pull-down menu, and then select the Customize menu item from the pull-down menu. As shown in the following illustration, selecting the Customize menu item will display the Customize dialog box.

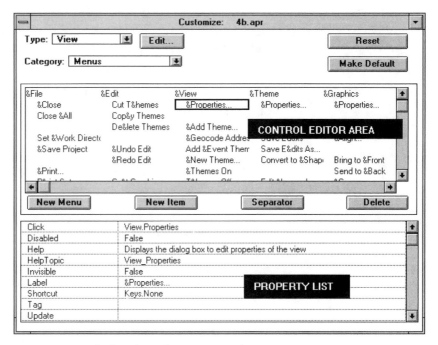

Customize dialog box for a view document.

Each ArcView document type has its own menu bar. Select the menu bar that you want to customize by choosing the ArcView document from the Type option at the top of the Customize dialog box. The Category option immediately below must be set to Menu. The current menu specification is displayed in the Control Editor area, with menu options in the top row. The menu items pertaining to each option are indented and appear below the option name.

Object Model for the Menu System

A menu system is created or modified either interactively through the customize dialog box, or dynamically by using Avenue scripts. You can use Avenue to manipulate a menu system while your application is executing. In order to understand how an Avenue script accesses your application's menu, you should first understand the object model for the menu system shown below.

Object model for the menu system.

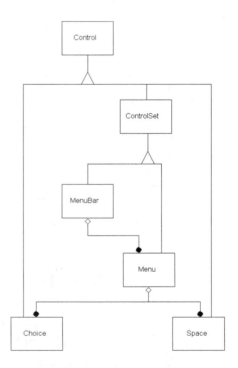

Control is an abstract class, encapsulating attributes and operations common to various user interface control objects. Interface controls may have a collection set. Menus, for example, have the menu bar as their collection set. ControlSet is another abstract class that defines common attributes and operations of such collection sets.

The previous object model diagram reveals that each menu bar is assembled from zero or more menu objects. A menu object appears on the menu bar and manages a collection of menu items and separator lines. Menu items are modeled as objects of the Choice class and separator lines are modeled as objects of the Space class.

Adding and Organizing Menu Options and Items

To customize interactively, select a menu option from the top row of the Control Editor area by clicking on it. A frame appears around the name to identify the selected option. Create a new menu by clicking on the New Menu button. Any new menu will be created to the right of the selected one. To place a menu option at the beginning of the menu bar, create it and then move it to the beginning. You can move a menu option and its associated items by dragging the option to a new location.

There is no limit on the number of menu options; Arc-View creates a second line for the menu bar if all options do not fit on one line. However, too many menu options could reduce readability and intuitiveness.

✓ **TIP:** *Limit the number of menu options to seven, and never create more than 12 for a single menu.*

Menu items are manipulated in much the same way as menu options. Create a new menu item by first selecting a menu option and then clicking the pointer on the New Item button. The new item will appear as the last entry on the option's list of items. If you select an existing menu item before creating a new one, the new menu item will appear after the existing item. You can move the items around within an option by dragging and dropping, but you cannot drag an item out of its menu option.

You might consider using a separator between groups of items to improve readability. To place a separator after an item, select that item and click the Separator button.

Because ArcView customization does not allow more than one tier of menus, menu items cannot be used to open a second list of menu items. Keep this limitation in mind when designing a menu system.

ArcView initially assigns *Menu* and *Item* as the names of created options and items, respectively. The defaults should be changed to more meaningful names related to the purpose or action of the menu. To change a name, double-click on the Label line inside the Properties list. A Label dialog box, shown in the following illustration, accepts the new name. There is no limit to the number of characters for the name, and all printable characters, including spaces, are acceptable. Lengthy names, however, make the application unfriendly.

Label dialog box.

✓ **TIP:** *Use a singular noun, such as File or View, to name menu options. Make all item options either a verb or a noun, such as Copy or Tile. Add an ellipsis (...) to the name of a menu item if its selection results in a dialog box.*

A menu system can also be created dynamically with Avenue. In the following code segment a new menu option with several menu items and a separator line is created for view documents.

 04-01.AVE

```
' This script assumes that a view document is active.
myViewGUI = av.GetActiveGUI
myMenuBar = myViewGUI.GetMenuBar
' Create a menu option to add to the menu bar.
myMenu = menu.Make
myMenu.SetLabel ("Index")
' Add the menu to the menu bar after the second
' existing menu option.
myMenuBar.Add (myMenu, 1)
' Create two menu items and add them to the
' menu option with a separator line.
myChoice1 = Choice.Make
myChoice1.SetLabel ("Zoom to last grid")
myChoice2 = Choice.Make
myChoice2.SetLabel ("Zoom to a grid...")
myMenu.Add (myChoice1, 0)
myMenu.Add (Space.Make, 1)
myMenu.Add (myChoice2, 2)
```

The Add request requires two parameters. The first parameter is the control you are adding to your control set, and the second parameter specifies the control's position. If you want to place a control at the beginning of the set, use -1 as the position. When you use the position 0 or higher, the new control is placed to the right of the existing control at the specified position. For example, specifying position 1 will place the new control as the third item. Using a position value higher than the number of existing controls places the new control at the end.

Executing a Task Through a Menu

A task is executed through a menu item by associating a script to the item. You can use an existing system script or write your own. Chapters 2 and 3 explain how to write new scripts.

Script Manager dialog box.

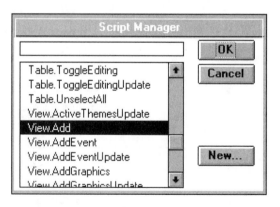

First, select the desired menu item. In the Properties list of the Customize dialog box, double-click on the Click property line to display the Script Manager dialog box. In this dialog box, shown in the following figure, available scripts are listed. The list includes both system and programmer-created scripts. Select the desired script from the list and click the pointer on the OK button. The name of the selected script appears in the Properties list of the menu item. You can review the contents of system scripts or create new ones based on existing scripts. This process is discussed in Chapter 2. You can remove a script from this property by selecting the property and pressing the <Del> key.

Disabling and Hiding a Menu Item

Under certain circumstances your application may require that specific menu options or items be unavailable for selection. In this case, you can either disable the item or make it invisible. A disabled item is displayed in gray but cannot be selected, while an invisible item is not displayed. Menu options or items are automatically rearranged to fill in the gap resulting from an invisible menu. The following figure shows the display change when Delete is disabled, and Window and Import are made invisible.

*Disabling and making
invisible menu
options and items.*

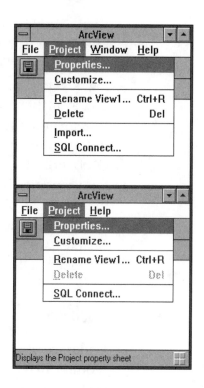

When an option is not accessible, its items also become inaccessible. Select the option or item you wish to disable or make invisible from the Control Editor area of the Customize dialog box. The Disabled and Invisible property lines in the Property list indicate the status of the menu option or item. A false value for the Disabled property means that the menu option or item is not disabled and can be selected. A false value for the Invisible property means that the menu option or item is displayed.

The value of a Disabled or Invisible property can be changed by double-clicking its property line, which toggles between true and false. The "Event Programming" section in this chapter explains how to change these properties with Avenue scripts.

Assigning Access and Shortcut Keys

Menu options can be opened with the keyboard instead of the mouse upon defining access keys. An access key is a selected character from an option's name. When the user holds the <Alt> key down and presses the access key, that menu option opens. Access keys are underlined when displayed on the menu bar.

Assign an access key by placing an ampersand before the desired character. For instance, *&Project* is assigned as the access key, so pressing <Alt+P> opens the Project menu option. Access keys are not case sensitive and can be any printable character. Avoid using a space for the access key as it may interfere with the reserved keys for a windowing system. If you need the ampersand displayed, precede it with a backslash. For instance, \ *&Left&Right* results in a menu label of *&LeftRight* with *R* as the access key. The ampersands following the access keys do not need the backslash. For instance, *&Left&RIGHT* results in a menu label of *Left&Right* with *L* as the access key.

Menu items can also have access keys. These are assigned in the same manner as for menu options. Using menu item access keys, however, differs from using keys to open a menu option. Once the user opens the list of menu items, simply pressing the character assigned as the access key opens the chosen item.

Access keys may be duplicated across menu options or among menu items. However, only the first option or item that was assigned a particular access key will always be selected. Duplicating access keys is not desirable in a well-designed application.

The user should be able to access frequently used operations (such as Save or Tile) directly, rather than always having to pass through the respective menu options. You can provide direct access to any menu item by assigning a shortcut property.

Select the desired menu item from the Control Editor area of the Customize dialog box. Double-click on the Shortcut property line in the Properties list. The Picker dialog box, shown in the following illustration, appears with a list of available keys. Select the desired key and click the OK button. The selected key names will appear in the Shortcut property line. Users can execute that menu item by pressing the assigned shortcut key or key combination. You can remove a shortcut by selecting Keys.None from the Picker list box.

Picker dialog box with a list of keys.

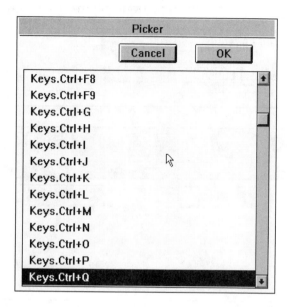

> ✓ **TIP:** *Use <Shift> or <Ctrl> in shortcut key combinations to reduce accidental executions of a menu item.*

Creating Buttons

Activate the Project window by clicking the pointer on its title bar. Select the Project menu option from the Project window control bar to display a pull-down menu. From the pull-down menu, select the Customize menu item to display the Customize dialog box shown in the following illustration.

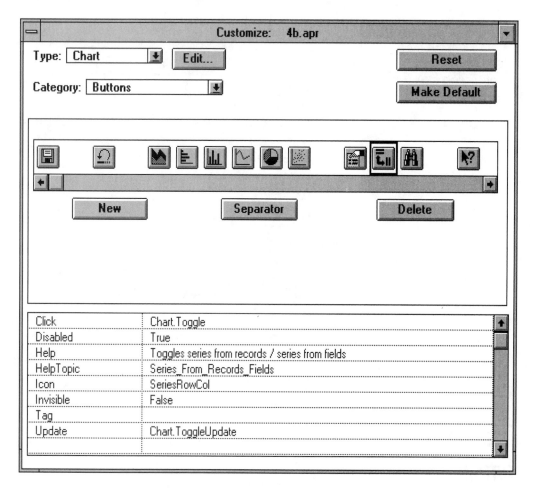

Customize dialog box for a chart document.

Each ArcView document type has its own button bar. Select the button bar you want to customize by choosing the ArcView document from the Type option. The Category option must be set to Buttons. The current button specification is displayed in the Control Editor area.

Object Model for Buttons

You can also manipulate buttons dynamically through Avenue scripts. Enabling and disabling buttons are the most common reasons for accessing buttons dynamically. Understanding the object model for buttons helps you in writing such Avenue scripts. An object model is shown in the following diagram.

Object model for buttons.

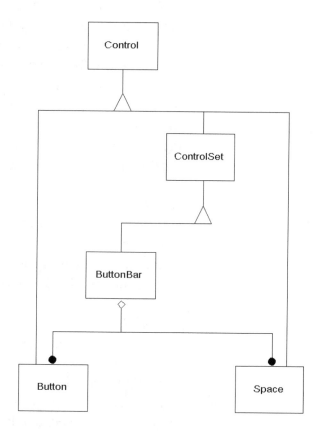

The object model diagram reveals that the button bar is a specialized class of ControlSet, which is an abstract class. Each button bar is optionally composed of one or more buttons and spaces. Button and Space are specialized classes of Control, which is also an abstract class.

Adding and Organizing Buttons

In the Control Editor area, click on a button in order to select it. A frame appears around the selected button. To create a new button, click on the New button. The new button will be created to the right of the selected button. You can move a button by dragging it to a new location. To place a button at the beginning of the bar, create it and move it to the beginning. If you drag the button and drop it on top of an existing one, ArcView places the new button to the left of the existing one. The Control Editor area is expanded by maximizing the Customize dialog box.

Although there is no limit to the number of buttons you can create, ArcView will display only a single line of buttons. Display resolution determines the number of buttons that can fit on the line. Too many buttons may reduce readability and intuitiveness.

✓ **TIP:** *Limit the number of buttons to 14 and never create more than 24.*

You might consider using a separator between groups of buttons to improve readability. To place a separator after a button, select that button and click on the Separator button.

ArcView creates buttons with a blank default icon. Because buttons cannot be labeled, it is important that the icon be representative of the button's action. To assign an icon to a button, select the button from the Control Editor area. Then double-click the pointer on the Icon line in the Properties list. The Icon Manager dialog box appears with a list of icons, as shown in the following illustration. The list shows the icon name and shape. Select an icon and click on the OK button. The icon will appear in the Properties list and on the button. Use the Load button on the Icon Manager dialog box to add your own icons to the list.

Icon Manager dialog box with a list of icons.

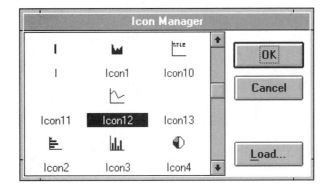

Loading a New Icon

ArcView maintains an extensive set of icons that you can select for customized buttons. However, you can create new icons and add them to your ArcView project. ArcView can accept icons in the following formats.

❒ GIF

❒ MacPaint

❐ NextpertObject Image

❐ SunRaster

❐ TIFF

❐ Windows Bitmap

❐ X Bitmap

Create your icon using drawing software. Limit the image size to 20 x 17 pixels for buttons. Store the image in one of the supported file formats. On the Icon Manager dialog box select the Load button to display the Load Icon dialog box, and then locate and select your image file. The image is added to the top of Icon Manager and the file name becomes the icon's name.

The new icon is stored with the project file. You must write an Avenue script to rename or delete the icon. The following code deletes an icon named *myicon.bmp* from the list of icons.

CODE   04-02.AVE

```
iconList = IconMgr.GetIcons
myIcon = nil
for each anIcon in iconList
    if (anIcon.GetName = "myicon.bmp") then
      myIcon = anIcon
      break
    end
end
iconList.RemoveObj (myIcon)
```

Executing a Task Through a Button

A task can be executed through a button by associating a script to the button. You can use an existing system script, or write your own.

Select the button to be affected from the Control Editor area of the Customize dialog box. Inside the Properties list, double-click on the Click property line. The Script Manager dialog box appears with a list of available scripts. The list includes both system and programmer-created scripts, as shown in the following illustration. Select the desired script from the list and click the pointer on the OK button. The selected script's name appears in the button Properties list.

Script Manager dialog box.

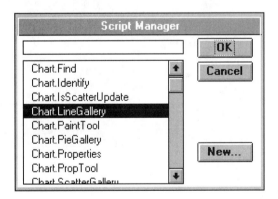

✓ **TIP:** *To find your script faster, use the text box on the Script Manager dialog box. For example, type VIEW* in the text box, followed by the <Enter> key, to limit the list to the script names that start with VIEW.*

Disabling and Hiding Buttons

Under certain circumstances, your application may require that a specific button be unavailable for selection. You can either disable the button or make it invisible. A disabled button is displayed in gray but cannot be selected, while an invisible button is not displayed. Bbuttonuttons are not rearranged to fill in the gap caused by an invisible button. The following illustration shows the change in the display when the Undo button is made invisible.

Before and after making the Undo button invisible.

Select the button you wish to disable or make invisible from the Control Editor area of the Customize dialog box. The Disabled and Invisible property lines in the Properties list indicate the state of that button. A false value for the Disabled property means that the menu is not disabled and can be selected. A false value for the Invisible property means that the menu is displayed. Change the value of the Disabled or Invisible property by double-clicking its property line, which toggles between true and false.

Buttons can also be disabled or made invisible through an Avenue script. The following code segment demonstrates this process.

 04-03.AVE

```
' This script is associated with the update property
' of a button for statistical analysis on the selected
' numeric field.
theTable = av.GetActiveDoc
aField = theTable.GetActiveField
if (nil = aField) then
    Self.SetEnabled (False)
elseif ( aField.IsTypeNumber ) then
    Self.SetEnabled (True)
else
    Self.SetEnabled (False)
end
```

Adding a Help Line to Buttons

You can provide a short description, or help line, for every button. The help line will appear on the status bar whenever the user places the pointer on a button. This is a very helpful feature because users often forget the actions button icons represent. Take advantage of it. The following figure shows the location of the status bar.

Status bar.

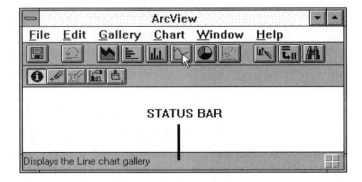

To define a help line, select the button to be affected from the Control Editor area. To display the Help dialog box, double-click on the Help property line inside the Properties list, as shown in the following illustration.

Although there is no limit on the number of characters, the help line should be short enough to read quickly. If the help line extends beyond the status bar, it will be truncated. Display resolution and default font determine the number of characters that can fit on the status bar. After typing the help line, click the pointer on the OK button. The help line text will appear in the Properties list.

Help dialog box.

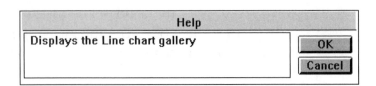

✓ **TIP:** *The text for a button's help line should be a complete sentence, and preferably less than 14 words. Avoid using more than 24 words.*

ArcView can show tool tips in Windows 95 and Windows NT environments. Tool tips are also known as balloon help or fly-by help. They appear when the mouse rests on a button. The string for the tip is provided in the help property using the following syntax.

```
tool tip//help string
```

Two forward slashes separate the strings for tool tip and status line help. If you provide only one string, ArcView uses it for the status line help. Use a very short string for the tool tip and capitalize the first letter of each word.

✓ **TIP:** *Use the name of the control as the tool tip string.*

Creating Tool Buttons

To create a tool button, make the Project window active by clicking on its title bar. From the control bar of the Project window, select the Project menu option to display a pull-down menu. Select the Customize menu item from the pull-down menu to access the Customize dialog box. Another way to open the Customize dialog box is to double-click on a blank area of the button bar or tool bar.

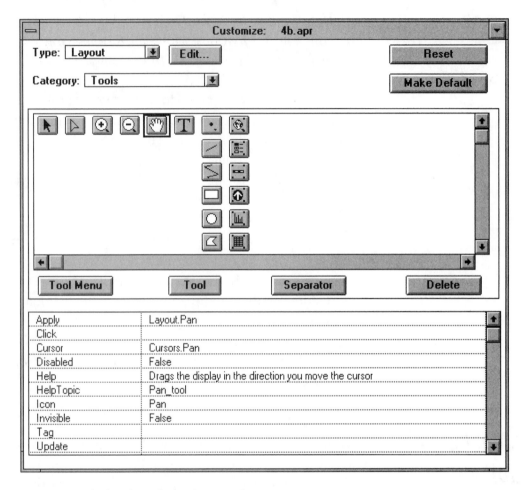

Customize dialog box for a layout document.

Each ArcView document type has its own tool bar. Select the tool bar that you want to customize by choosing the ArcView document from the Type option. The Category option must be set to Tools. The current tool specification is displayed in the Control Editor area.

Similar to other controls, tools can be dynamically changed through Avenue scripts. The following diagram shows the object model as related to tools. Understanding this diagram can help you when writing Avenue code for tools.

Object model for tools.

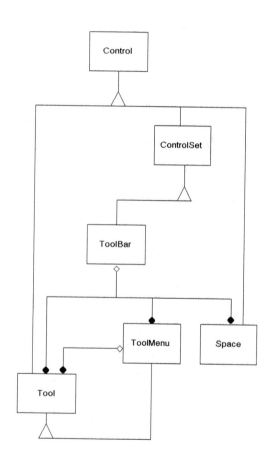

The previous diagram demonstrates placing tools, tool menus, and spaces on a tool bar. The tool menu contains multiple tools.

Adding and Organizing Tool Buttons

Select a tool from inside the Control Editor area. A frame will appear around the selected tool button. To create a new tool button, click on the Tool button and the new tool will be created to the right of the selected button. To place a tool button at the beginning of the bar, create it and then move it to the beginning. You can move a button by dragging it to its new location. If you drag the button and drop it on top of an existing one, ArcView places the new button to the left of the existing one. The Control Editor area is expanded by maximizing the Customize dialog box.

There is no limit on the number of tool buttons you can create. However, ArcView displays all buttons in a single line, and display resolution determines the number of buttons that can fit on the line. Too many buttons could reduce readability and intuitiveness.

✓ **TIP:** *Limit the number of tool buttons to 14, and never create more than 24.*

ArcView creates tools with a blank default icon. Because tool buttons cannot be labeled, it is important that the icon represent the button's action. To assign an icon to a tool button, select the button from the Control Editor area. Then, double-click on the Icon line in the Properties list. The Icon Manager dialog box appears with a list of icons, as shown in the following figure. The list shows the icon name and shape. Select an icon and click on the OK button. The icon will appear in the Properties list and on the button.

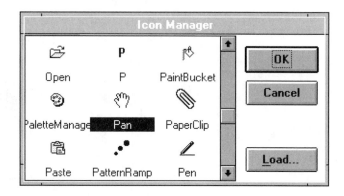

Icon Manager dialog box with a list of icons.

To learn how to add your own icon, see the "Loading a New Icon" section in this chapter.

✓ **TIP:** *Tool icons are 18 by 15 pixels.*

You can create a tool menu by clicking on the Tool Menu button. Subsequently, you can add tools to the tool menu by clicking on the Tool button.

Executing a Task Sequence Through a Tool

A task sequence can be executed through a tool button by associating a script to the button. You can use an existing system script, or you can write your own.

Select the button to be affected from the Control Editor area of the Customize dialog box. Inside the Properties list, double-click on the Apply property line to display the Script Manager dialog box. Available scripts are listed in this dialog box, as shown in the following illustration. The list includes both system and programmer-created scripts. Select the desired script from the list and click the pointer on the OK button. The name of the selected script will appear in the tool button's Properties list.

Script Manager dialog box.

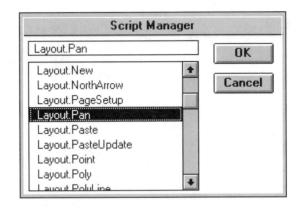

Associating a Cursor Shape to a Tool

When selecting a tool, the user expects to apply the tool's actions by clicking or dragging the pointer inside the document window. For instance, after selecting the Zoomin tool, the user expects to zoom in to the area clicked on by the pointer. The user may repeat the tool's action as long as another tool is not selected. Consequently, it is important to change the shape of the cursor to indicate which tool is active.

You can associate a cursor shape to a tool button by selecting the button to be affected from the Control Editor area. Double-click on the Cursor property line inside the Properties list and the Picker dialog box will appear with a list of cursors. The graphical shape and name of each available cursor are displayed in this list.

Select the desired shape and click the OK button. The cursor name will appear in the Properties list.

Picker dialog box with cursors.

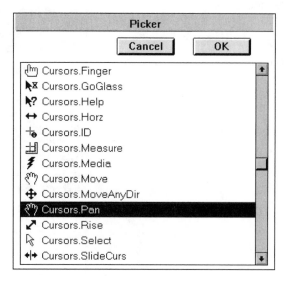

✓ **TIP:** *To aid user recognition, the cursor's graphical shape should match or be conceptually related to the tool button's icon.*

Disabling and Hiding a Tool

Your application may sometimes require that certain tool buttons be unavailable for selection. You can either disable tool buttons or make them invisible. A disabled button is displayed in gray to indicate that it cannot be selected. An invisible button is not displayed. Tool buttons are not rearranged to fill in the gap caused by an invisible tool. The following figure shows the changes in the display when tools are disabled and made invisible.

*Before and after
disabling and
making invisible tool
buttons.*

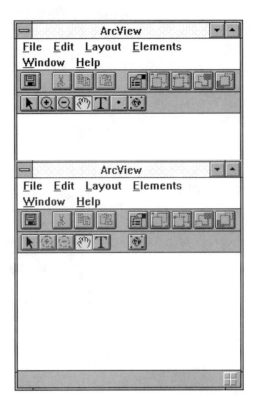

Select the tool button you wish to disable or make invisible from the Control Editor area of the Customize dialog box. The Disabled and Invisible property lines in the Properties list indicate the state of the button. A False value for the Disabled property means that the button is not disabled and can be selected. A False value for the Invisible property means that the button is displayed. Double-click on the Disabled or Invisible property lines to toggle between true and false. The "Event Programming" section in this chapter explains how to change these properties through Avenue scripts.

Adding a Help Line to Tools

You should take advantage of the opportunity to provide a short description or help line for each tool button. This is a very helpful feature because users often forget which action a button's icon represents. The help line will appear on the status bar, as shown below, whenever the user places the pointer on a tool.

Status bar.

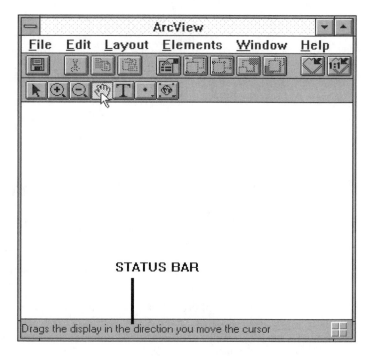

To define a help line, select the tool button to be affected from the Control Editor area. Double-click on the Help property line inside the Properties list, and a help dialog box, as shown below, accepts the help line.

Although there is no limit on the number of characters in a help line, you should keep it short for easy reading. If the help line extends beyond the status bar, it will be truncated. Display resolution and default font determine the number of characters that can fit on the status bar. Click the pointer on the OK button after typing the help line, and the help line text will appear in the Properties list.

Help dialog box.

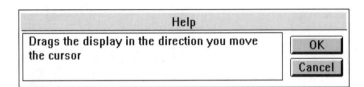

✓ **TIP:** *Text for a tool button's help line should be a complete sentence and preferably less than 14 words. Avoid using more than 24 words.*

ArcView can show tool tips in Windows 95 and Windows NT environments. Tool tips are also known as balloon help or fly-by help. They appear when the mouse rests on a tool. The string for the tip is provided in the help property using the following syntax.

```
tool tip//help string
```

Two forward slashes separate the strings for tool tip and status line help. If you provide only one string, ArcView uses it for the status line help. Use a very short string for the tool tip and capitalize the first letter of each word.

✓ **TIP:** *Use the name of the control as the tool tip string.*

Creating Pop-up Menus

Pop-up menus can be created or changed through the customize dialog box. Open the Customize dialog box by double-clicking on a blank area of the button bar or tool bar, and select the pop-ups from the category list. Clicking the New Pop-up button creates new pop-up menus. Although you can create many pop-up menus in a set, only one can be displayed. By default, the leftmost pop-up menu is always active, but you can change that through Avenue. The following code segment shows how to make a pop-up menu active.

```
aPopupSet = av.FindGUI ("VIEW").GetPopups
aPopupSet.SetActive (aPopupSet.FindByLabel("MYPOPUP"))
```

Add menu items to your pop-up menu by clicking on the New Item button in the Customize dialog box. Pop-up menu and item properties are similar to those explained in the "Creating a Menu System" section of this chapter.

Customizing the Project Window

You can customize the project window by clicking on the Edit button of the Customize dialog box to display the Customize Types dialog box. Use this dialog box, shown below, to customize the project window.

Customize Types dialog box.

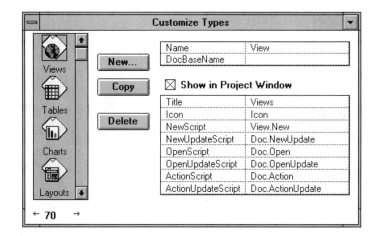

You can create new groups based on the following existing types: view, table, layout, chart, script editor, application, and project. For example, you could create a new group based on the view document and label it Project Index, as shown in the following figure. Then you can place index views in that group.

A customized project window.

Several properties must be defined when creating a new document group. The following figure shows these properties for a new type named *Index*.

Customize Types dialog box.

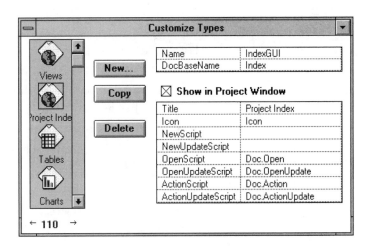

The Name property assigns a name for the DocGUI
object representing the new group's interface. When
you need to access this DocGUI you would use its
name as shown in the following code segment.

```
myIndexGUI = av.FindGUI("IndexGUI)
```

The Document Base Name property establishes the
default name for the new documents of this group. For
instance, when it is set to Index, new documents are
named *Index1*, *Index2*, and so forth. You must check
the Shown in Project Window check box in order to
see the new group. To place the new group, drag its
icon.

The Title property sets the icon's label on the project
window. If your label is too long, and therefore truncated, increase the number of pixels that set the width
of the left pane on the project window. The icon property is set to Icon to represent the default icon for the
document type. You can change it by double-clicking
on that property line.

The remaining properties associate the click and
update scripts to the three buttons on the project window. Although you can change the label of all three
buttons, the first two are generally used for creating
new documents and opening existing ones. The third
button is known as the action button and is used for
printing views and layout, adding tables, and running
scripts. Its label is set by the script associated with its
update event. If you do not want a button to be visible,
remove its click and update scripts. To remove a

script, highlight its property line and press the <Del> or <Backspace> key.

Start-up and Shutdown Scripts

Many applications can benefit from the ability to run a script based on the occurrences of an event, such as starting or ending a project. You can associate start-up and shutdown scripts to your project. Start-up scripts are useful for setup or verification of the application environment, while shutdown scripts are generally used for cleanup or validation.

Start-up and shutdown scripts are attached in the Properties dialog box. The following illustration shows the Properties sheet for a project. Open the Properties dialog box by selecting the Project menu option. Choose the Properties menu item from the option's pull-down menu. Enter the script name in the Properties dialog box.

Project Properties dialog box.

You can use either a system script or a script that you have created. If you cannot recall the script's name, click the pointer on the Loadscript icon. The Script Manager dialog box appears with a list of available scripts from which to choose.

ArcView documents also have the ability to run a script when they are open or closed, in addition to when their interface changes. You can associate open, closed, and updated scripts through Avenue as shown in the following code segment.

```
myIndex.SetOpenScript ("index_open")
myIndex.SetCloseScript ("index_close")
myIndex.SetUpdateScript ("index_update")
```

Saving the Customized Interface

Changes made to the interface are stored with the project. Every time you save an ArcView project, the changes to the interface, along with the scripts and documents, are saved to the ArcView project file. You can distribute the application by distributing the project file. If you want to share the interface changes with others, or use the changes in new projects, set the current project as the default project file. In this manner you can create a personalized environment. ArcView has two default levels: system and user level. User level defaults supersede system level defaults. Defaults are stored in a project file named *default.apr.*

Every time ArcView begins, it first reads the system default file. The system default file establishes Arc-View's non-customized interface. Next, ArcView reads the user default file if available. The user default file may contain a customized interface that overwrites those from the system default file. Finally, when opening a project file, its customized interface overwrites the user default file interface.

Save a customized interface to the user-level default file by opening the Customize dialog box and clicking the pointer on the Make Default button. The Customize dialog box is accessed by making the Project window active, selecting the Project menu option, and then picking the Customize menu item. A user-level *default.apr* file is then created in one of the following platform-dependent directories.

❐ %HOME% for Microsoft Windows family

❐ ArcView application folder for Macintosh

❐ $HOME for UNIX

❐ SYS$LOGIN for Open VMS

Do not overwrite the system default file, *default.apr,* in the *etc* directory with the user level *default.apr.* The system default file contains all scripts and user interface controls while the user default file contains only the scripts and controls different from the system default.

✓ **TIP:** *Existing projects are not affected by new defaults. If you press the Reset button in the Customize dialog box, your interface becomes the current system default setting.*

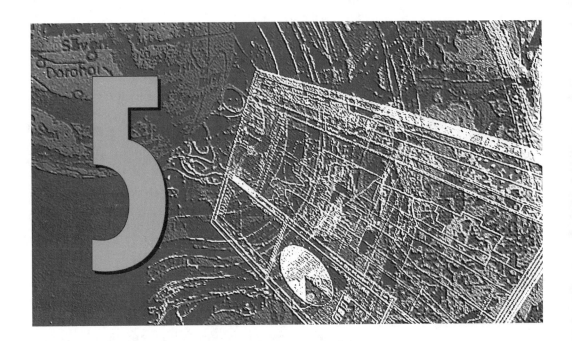

Programming the Project Window

A collection of associated documents in ArcView is called a *project*. The project window organizes and displays the names of the documents. Although your application will generally access associated documents more frequently than the project window, the project window has a set of features that can be useful in all applications.

This chapter provides an in-depth examination of an application for estimating damages caused by a brush fire. When working with this application, the user loads coverages of planimetric features along with the final boundary of the damaged area to estimate total damages to property and infrastructure.

The chapter ends with a tutorial that walks you through the process of creating a simple application.

Setting Up an Application

How might typical users start your application? They could double-click on an icon or type the application name on a command line. Whatever the chosen process, it must be easy to remember and simple to perform. This section presents an alternative, shown in the following illustration, that modifies ArcView's interface with menu items to start an application.

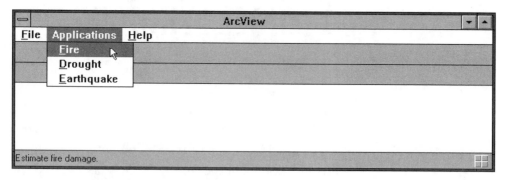

Customized ArcView interface.

A Customized ArcView User Interface

The purpose of the above interface is to provide the user with a quick and convenient way to begin an application. Listed below are the steps required in customizing the interface as discussed in Chapter 4.

❏ Start a new project.

❏ Make the Project window active and select the Customize menu item from the Project menu option.

❏ Set the category option to Menu, and the type option to Appl (application).

❏ In the File menu options, make the New Project menu item invisible.

✓ *TIP: When a control interface such as a menu option or tool button is not needed, make it invisible instead of deleting it. You may need it later.*

Create a menu item for each of your applications. The scripts initiate applications by opening designated projects. For instance, the script below starts the Fire application. The next figure shows the properties of the Fire menu item. For each menu item, you must create an Avenue script. Once everything is in place, click on the Make Default button and close the Customize dialog box.

Fire menu
item properties.

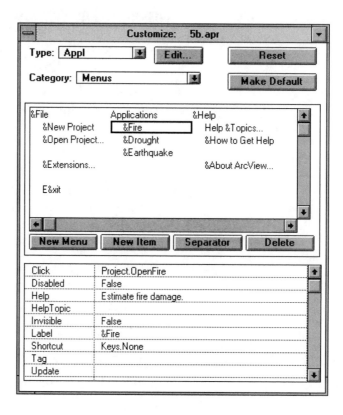

The following script starts a project named *_FIRE*. The purpose of the underbar is to create a unique file name; it is unlikely that users will begin project file names with an underbar. Users should save projects with new names. For example, assume that on three occasions a user launches the _FIRE application by starting the project of the same name. The application is activated to estimate the damage from a specific fire, and results are saved in projects named *CALAV,* *SHASTA,* and *ELDORADO,* respectively. The _FIRE project file remains free of data and can be used at any time for a new fire, while *CALAV, SHASTA,* and *ELDO-RADO* hold information on individual fires.

```
' Start the _FIRE project.
projectFile = FileName.Make ("_FIRE.APR")
Project.Open (projectFile)
```

Once a specific project file such as *SHASTA* is created, the user will open that project directly from the File menu option. For instance, if the user wants to update the *SHASTA* project, ArcView would be started, and the Open Project menu item selected from the File option. The user would then select *SHASTA.APR* from the file window.

➥ **NOTE:** *ArcView opens an empty project every time it is started. In order to see the Applications menu option just created, you must close the empty project.*

Application Start-up Script

The best time to enforce saving the project with a new name is when ArcView opens the *_FIRE* project. The following Avenue script should be placed in the Startup property of the *_FIRE* application to force a save with a new name.

CODE 05-01.AVE

```
' Forcing Save As at the start-up.
theProject = av.GetProject
defaultName = FileName.GetCWD.MakeTmp ("fire","apr")
' Open the file window to get a project name.
newProject = FileDialog.Put
(defaultName,"*.apr","Create a New Project")
if (nil <> newProject) then
   theProject.SetFileName (newProject)
   theProject.Save
else
   ' If the user clicked on Cancel, close the project.
   theProject.Close
end
```

Setting Project Parameters

To assign Avenue scripts to start-up and shutdown
events, use the Project Properties dialog box. To open
this dialog box, select the Properties menu item from
the Project menu option. The following figure shows
the Project Properties for the _FIRE_ application.

*Project Properties
dialog box.*

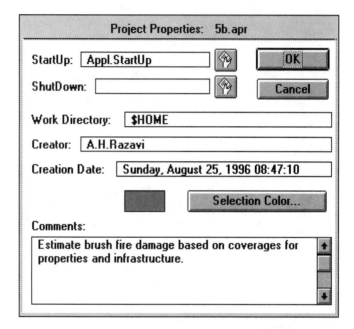

This dialog box can also store general information about
the project. The importance of documentation cannot be
overemphasized; key your name in the Creator field and
provide a short description under Comments.

Application Framework

An ArcView project is modeled by the instances of the Project class. You can better understand this class by reviewing the application framework. The following diagram shows the object model as it relates to the application framework.

Object model for the application framework.

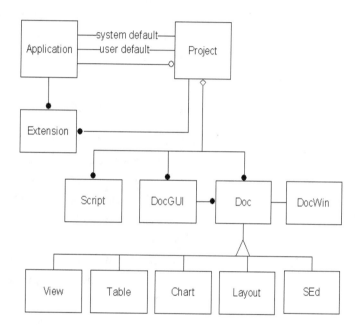

The only instance of the Application class is the Arc-View program, which can be accessed in Avenue through the keyword *av*. You can access almost all objects in ArcView from *av* because it contains all the objects that comprise ArcView. The default name for this object is ArcView, but you can change it by using the SetName request. The name appears in the Arc-View title bar.

The Extension class provides add-on capability to Arc-View. See Chapter 12 for a description.

The preceding diagram shows that a project consists of scripts, document GUIs, and documents. Doc is an abstract class for various ArcView documents. These documents allow interaction with data. For example, the table document provides access to tabular data. ArcView documents are specialized classes of Doc. The next section shows how you can access these documents through the Project object.

Finding a Document

The project window provides direct access to all documents associated with the project. The ArcView user can create, open, or delete documents with the project window. An application developer can use the project object in Avenue to access document objects. For instance, for a list of themes to appear in a view, the view object must be accessed. To access the view object, you can search for it in the current project object.

There are two ways to reference a document object. When the Avenue script is associated with a document through an interface control, you can ask for the active document. If the script is not associated with a document in this fashion, you can search for the document by name. The following code segment shows how to reference a document.

```
' Open the view named Final.
thisProject = av.GetProject
```

```
theView = thisProject.FindDoc ("Final")
theViewWindow = theView.GetWin
if (theViewWindow.IsOpen) then
    theViewWindow.Activate
else
    theViewWindow.Open
end
' Show or hide the legend for the current view.
' Associate this script with a button.
theView = av.GetActiveDoc
for each oneTheme in theView.GetThemes
    oneTheme.SetLegendVisible
    (Not(oneTheme.IsLegendVisible))
end
```

The following script shows how to create a list of views in a project.

05-02.AVE

```
docList = av.GetProject.GetDocs
viewList = {}
for each aDoc in docList
    if (aDoc.Is(View)) then
        viewList.Add (aDoc)
    end
end
MsgBox.List (viewList,"","Views")
```

Arranging Document Windows and Icons

During the course of an ArcView session, users could create or open so many documents that the ArcView window becomes cluttered. The Window menu option can be used to manually organize all the documents by arranging them as tiles or cascading them next to each other. Another option is to click on minimize buttons to reduce the documents to icons. These capabilities are also available for inclusion in your application through the following Avenue statements.

```
av.TileWindows
av.CascadeWindows
av.ArrangeIcons
```

In the following code segment, _FIRE_ opens several views and then arranges them in tile format.

```
' Open views of FIRE.
view3amWin.Open
view9amWin.Open
view3pmWin.Open
' Arrange views on the screen.
av.TileWindows
```

Displaying a Help Message

Window-based applications with graphical user interfaces (GUIs) rely on the user to direct the application in executing functions. This reliance on the user is the tradeoff for the flexibility offered by windows.

In contrast, a non-GUI application requests pertinent data from the user in a serial fashion. Failure to respond to some questions is allowed, but the application will not start computation until all questions have been displayed. If the application was equipped with a GUI, the user could indicate which data are available for input by selecting menu items or buttons, thereby avoiding the display of irrelevant questions. In addition, the user could click on a button at any time to start computation based on the available data.

In a GUI application, a short message indicating how to start, or what the next step is, can be of great help to the user. For instance, if _FIRE_ expects selection of a base coverage or an image file, the following message could be displayed on the status line.

Load a base map or image file by selecting the Load menu option.

The Avenue requests that handle messages appear below.

```
av.ShowMsg ("Message string")
av.ClearMsg
```

Messages appear in the status line. ArcView also uses the status line to display help lines for the interface items. When incorporating help messages into your application, keep in mind that ArcView messages are also displayed in the status line. Thus, if your application and ArcView act to display messages at the same moment, application-generated messages will be overridden by ArcView's.

One effective use of status line messages is to confirm process or task completion. In this case, the advantage of using the ShowMsg request over a message box request (MsgBox) is that ShowMsg does not require acknowledgment. For the same reason, however, you should not use the ShowMsg request to display critical information.

There is no limit on the number of characters in a status line message. However, ArcView will truncate messages longer than the status line. The + and ++ operators can be used to assemble the message string.

Displaying the Status Bar

When ArcView opens a project, it displays a status bar showing the progress of opening the project. A progress report is useful for application tasks that are usually time-consuming, and lack other means to show progress such as changes to a view. Another benefit of using a status bar is that the user can interrupt the process.

The Avenue requests that show a status bar and stop button appear below.

```
av.ShowStopButton
continued=av.SetStatus (aNumber)
av.ClearStatus
```

The parameter *aNumber* ranges from 0 to 100. The status bar is displayed at zero, and at 100 the status bar is cleared. Each SetStatus request moves the status bar to the location indicated by *aNumber*. The request returns a Boolean True value unless the user has clicked on the Stop button.

The following code segment is a simple example of how the status bar can be used to show a progress report.

```
theView = av.GetProject.FindDoc ("Final")
themesList = theView.GetThemes
' Start the status bar with stop button.
av.ShowMsg ("Computing damages...")
canceled = False
av.ShowStopButton
statusIndex = 0
av.SetStatus (statusIndex)
' Status bar reaches 100% when computation
' is performed for all themes.
```

CODE   05-03.AVE

```
themesCount = themesList.Count
statusIncrement = 100 / themesCount
' Application loops through all themes.
for each aTheme in themesList
   av.Run("Fire.ComputeDamage",aTheme)
   statusIndex = statusIndex + statusIncrement
   continued = Av.SetStatus (statusIndex)
   if (Not continued) then
     canceled = true
     break
  end
end
' Handle any interruption.
if (canceled) then
   av.ShowMsg ("Process interrupted.")
else
   av.ShowMsg ("Process completed.")
end
```

Tutorial: Creating an Application

This tutorial gives you a taste of Avenue programming by guiding you through the development of a simple application. The tutorial assumes that you are knowledgeable about ArcView, and do not need instruction on tasks such as saving a project or renaming a view.

For the practice application, the *states shape* file shipped with ArcView will be used. You should be able to locate the coverage in ArcView's *usa* directory. The application provides menu options to select one or more states, and enables the user to build a population density chart, print the chart data to a file, and compose a map.

Development Overview

When designing window-based applications, you must consider the following questions: Which objects are needed by the application? How does the application respond to events?

You will work with ArcView objects, such as menu items, charts, and graphic elements. Events are actions taken by the user, such as clicking the mouse on a button or opening a view document. The application responds to events through the Avenue scripts you write. The combination of ArcView objects and Avenue scripts make up an application.

The objects listed below are required for the application tutorial.

❐ Menu options and items

❐ A view document with the *States* theme

❐ The theme's attribute table

❐ A chart document

❐ A layout document

Menu Items

The menu items used here are Select States, View Chart, Compose Map, and Report.

Select States. Opens the view document with the *States* theme. The program verifies that the view extent is set to the theme, and clears any other selected state.

View Chart. Opens a predefined chart document and redraws its contents to match the current selection. It must be disabled if no state has been selected.

Compose Map. Opens a layout document to allow the user to compose a map of the states.

Report. Generates a delimited text file from selected records. If no records are selected, all records are written.

Development Stages

The practice application is created in the following stages.

Stage 1. Create required documents, such as the view or the chart.

Stage 2. Customize the interface by adding the menu option and items.

Stage 3. Develop Avenue scripts that respond to the selection of menu items.

Stage 1: Application Documents

In this section, you will create a project file, a new view document, and join two tables.

1. Create the project file by opening a new project and saving it as *GUIDE.*

2. Create a new view document and rename it *USA_View.* Access the Properties dialog box of this document to rename it.

3. Load the *States* theme into the *USA_View* document.

4. Make the *States* theme active, and select Table from the Theme menu option. This action loads and opens the *Attributes of States* table.

5. Close the table and view documents.

6. Click on the Table icon in the project window. Load the *Stdemog.dbf* table by selecting the Project menu option, then the Add Table menu item. This file is in the tables subdirectory.

7. Join the *Stdemog* table containing demographic information with the *Attributes of States* table. The result should go to the latter table. Join the two by opening both tables and selecting the *State_name* field in each table. Then, while the *Attributes of States* table is active, select the Join menu item from the Table menu option.

8. Keep the *Attributes of States* table open and active. Create a chart by selecting the Table menu option and then the Chart menu item. In the Chart Properties dialog box, add *Pop90_sqmi* to the Groups list, and place *State_name* in the Label series using field.

9. Rename the newly created chart *Density_Chart.*

10. Save the project.

Stage 2: Customize the Interface

In this section, you will customize the interface using techniques discussed in Chapter 4.

1. Open the Customize dialog box by double-clicking on the button bar.

2. Set the Type option to Project, and the Category option to Menus.

3. Click on the New Menu button to create a new menu option. Verify that it is selected. Change its name by double-clicking on the Label property. When the Label dialog box appears, key in *&Guide* and click on the OK button.

4. Add a menu item to the Guide menu option by clicking the mouse on the New Item button. Change the name of the new item to *&Select States* by double-clicking on its Label property. Add shortcut keys by double-clicking on the Shortcut property to display the Picker dialog box. From the Picker dialog box, select Keys.Shift+F1 and click on OK. Finally, add a short help text. Double-click on the Help property to open the Help dialog box. In the Help dialog box, key in *Select states from the USA View document;* click on OK. The following figure shows the Customize dialog box.

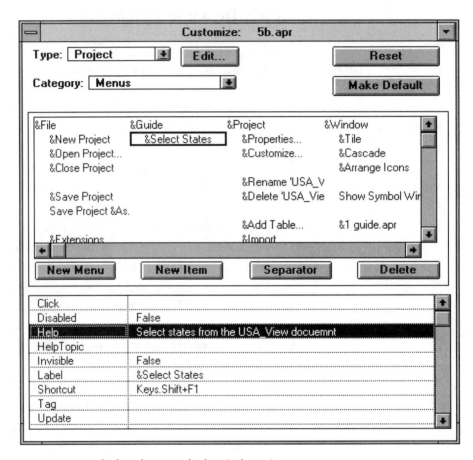

Customize dialog box with the Select States menu item.

5. Follow the specifications below when adding three additional menu items to the Guide menu option. See the instructions in step 4 above.

```
Label: View &Chart
Shortcut: Keys.Shift+F2
Help: View density chart for the selected states
Label: Compose &Map
Shortcut: Keys.Shift+F3
Help: Open a layout document
Label: &Report...
Shortcut: Keys.Shift+F4
Help: Print data to a file
```

6. In the Customize dialog box, change the Type option to View. Maintain the Category at Menus. Add the Guide menu option with the View Chart and Compose Map menu items according to steps 4 and 5 above. The following figure shows the Customize dialog box for this step. The user interface for view documents has a Graphics menu option with *G* as the access key. Change the Graphics access key to lower case *r* to avoid confusion with the *G* access key in the Guide menu option.

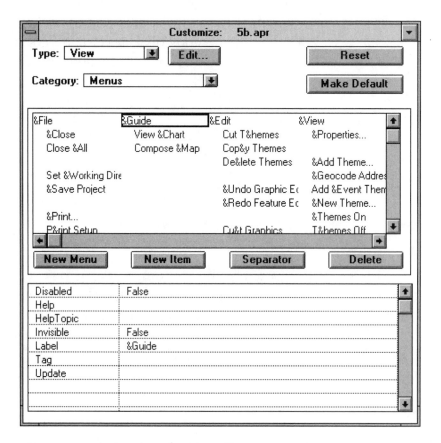

Customize dialog box for View Document.

7. In the Customize dialog box, maintain the Category option at Menus but change the Type option to Chart. Add the Guide menu option with the Select States, Compose Map, and Report menu items, according to steps 4 and 5.

8. The user interface for chart documents has a Gallery menu option with *G* as the access key. To avoid any confusion with the Guide menu option access key, change the Gallery option access key to lower case *y*. To change the access key, verify that the Type option is Chart, select the Gallery menu option, and double-click on the Label property. When the Label dialog box appears, change *&Gallery* to *Galler&y*.

9. Compare your user interface to the next three figures.

Project window user interface.

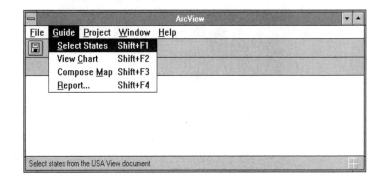

View document user interface.

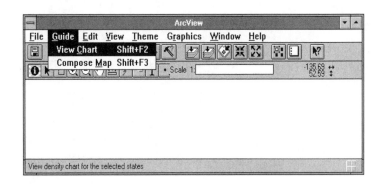

*Chart document
user interface.*

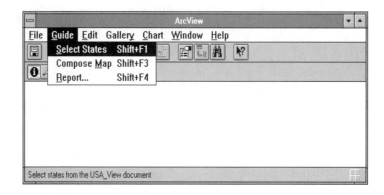

10. Save the project.

Stage 3: Avenue Scripts

Thus far, you have assembled the objects required by the practice application, as well as the necessary Arc-View documents and user interface controls. The next step is to design the application's response to events. Events are user selections of the menu items you just created. Responses are the actions that your application takes when the menu items are selected. For instance, when the user selects the View Chart menu item, the application should open the Density_Chart document created earlier.

You will need the following five Avenue scripts.

❐   Script name: Guide.SelectStates
    Purpose: Open the USA_View document.

❐   Script name: Guide.UpdateChart
    Purpose: Disable the View Chart menu item if no state is selected.

❐ Script name: Guide.ViewChart
Purpose: Open the Density_Chart document.

❐ Script name: Guide.Print
Purpose: Write the selected records to a comma-delimited text file.

❐ Script name: Guide.NewLayout
Purpose: Create a new layout document or open an existing layout document.

The project name *Guide* precedes all script names for two reasons. First, when you open the Script Manager dialog box to associate these scripts to respective menu items, menu items will be listed in groups by script, and thus are easy to find. Second, as you write numerous scripts in the future, including the project name as part of the script name can be used as a script management tool.

A script is created by opening the Script Editor. In the Project window, click on the Script's icon and select the New button to open the Script Editor. ArcView gives each new script a default script name, which is not usually descriptive of the script's contents. You should give the new script a more meaningful name. A script is renamed by opening its Properties dialog box. To open the Properties dialog box, select the Script menu option, then select the Properties menu item.

Guide.SelectState Script

Key in the following Avenue code and name it *Guide.SelectState*. Click on the Compile button. If there are no syntax errors, save your work. If the compiler locates errors, review your script for any deviation from the code.

05-04.AVE

```
' Guide.SelectState
' Open the USA_View document.
'
guideProject = av.GetProject
usaView = guideProject.FindDoc("USA_View")
'
' Stop if the view document does not exist.
if (nil = usaView) then
   MsgBox.Error
   ("USA_View document does not exist.","Guide")
   exit
end
'
' Open the view document.
usaViewWin = usaView.GetWin
if (usaViewWin.IsOpen.Not) then
   usaViewWin.Open
else
   usaViewWin.Activate
end
'
' Get the states theme object.
stateTheme = usaView.FindTheme ("States.shp")
'
' Stop if theme does not exist.
if (nil = stateTheme) then
   MsgBox.Error
   ("States theme does not exist.","Guide")
   exit
end
```

```
'
' Display the theme.
if (stateTheme.IsVisible.Not) then
    stateTheme.SetVisible (true)
end
'
' Clear any selected features.
stateTable = stateTheme.GetFTab
allSelected = stateTable.GetSelection
allSelected.ClearAll
stateTable.UpdateSelection
'
' Zoom to the extent of state theme.
If (stateTheme.IsActive.Not) then
    stateTheme.SetActive (true)
end
usaView.GetDisplay.SetExtent
(stateTheme.ReturnExtent.Scale(1.1))
```

Associate this script with the Select State menu item. Remember that the Select State menu item appears in more than one place. Open the Customize dialog box and click on the Select State menu item. Double-click the item's Click property to open the Script Manager dialog box, and select the Guide.SelectState script from the list.

Guide.UpdateChart Script

Use the following code to create a new script called *Guide.UpdateChart.* Compile the code and then associate it with the Update property of the View Chart menu item. Remember that this menu item appears in more than one place.

05-05.AVE

```
' Guide.UpdateChart
' The View Chart menu item should be disabled
' when no state has been selected.
'

guideProject = av.GetProject
usaView = guideProject.FindDoc ("USA_View")
'

' Keep menu item disabled if view does not exist.
if (nil = usaView) then
   SELF.SetEnabled (false)
   exit
end
'

' Count selected feature items.
stateTable = usaView.FindTheme("States.shp").GetFTab
selectedCount = stateTable.GetSelection.Count
'

' Disable or enable based on the count.
if (selectedCount > 0) then
   SELF.SetEnabled (true)
else
   SELF.SetEnabled (false)
end
```

Guide.ViewChart Script

The following script opens the *Density_Chart* document and redraws its contents. Name the script *Guide.ViewChart,* compile it, and associate it with the click property of all the View Chart menu items.

 05-06.AVE

```
' Guide.ViewChart
' Open the chart named Density_Chart.
'
guideProject = av.GetProject
densityChart = guideProject.FindDoc ("Density_Chart")
'
' Stop if the document does not exist.
if (nil = densityChart) then
   MsgBox.error
   ("Density_Chart document does not exist.","Guide")
   exit
end
'
' Open the document.
densityChartWin = densityChart.GetWin
if (densityChartWin.IsOpen.Not) then
   densityChartWin.Open
else
   densityChartWin.Activate
end
'
' Redraw the chart to accept new selection.
densityChartWin.Invalidate
```

Guide.Print Script

Associate the following script with the click property of the Report menu item. This script asks the user to supply a file name and then writes the selected records to the file.

05-07.AVE

```
' Guide.Print
' Write the selected records to a delimited
' text file; if none are selected, write all
' records.
'

' First, get the table object.
usaProject = av.GetProject
usaTable = usaProject.FindDoc
("Attributes of States.shp")
if (nil = usaTable) then
   MsgBox.Error
   ("Attributes of state table does not exist","Guide")
   exit
end
'

' Get an output file name.
aFileName = FileDialog.Put
("*.txt".AsFileName,"*.txt","Print to a file")
if (nil = aFileName) then
   ' Stop if user clicks on the Cancel button.
   exit
end
'

' Write to the file.
usaVTab = usaTable.GetVTab
usaVTabSelectionCount = usaVTab.GetSelection.Count
if (usaVTabSelectionCount = 0) then
   usaVTab.Export (aFileName,DText,false)
   MsgBox.Info
   ("All states written to"++aFileName.GetFullName,
   "Guide")
else
```

```
usaVTab.Export (aFileName,DText,true)
MsgBox.Info
("Selected states written to"++
aFileName.GetFullName,"Guide")
end
```

Guide.NewLayout Script

The script below asks the user for a layout name. If the layout exists, the script opens it; if it does not exist, the script creates the new layout. Associate this script with the click property of the Compose Map menu item.

CODE 05-08.AVE

```
' Guide.NewLayout
' Create a new layout document
' or open an existing one.
'
usaProject = av.GetProject
'
' Get a layout name from the user.
layoutName = MsgBox.Input
("Enter a layout name:","Guide","")
if (nil = layoutName) then
    ' Stop if the user clicked on the Cancel button.
    exit
end
aLayout = usaProject.FindDoc (layoutName)
if ((nil = aLayout)or
(aLayout.Is(Layout).Not)) then
    ' Create a new layout.
    aLayout = Layout.Make
    aLayWin = aLayout.GetWin
    aLayWin.Open
    aLayout.SetName (layoutName)
else
    ' Open the existing layout.
    aLayout.GetWin.Open
end
```

Testing the Application

The testing process for large applications often includes a test plan and document to guarantee a complete check. For small applications such as this one, a formal plan may not be necessary. However, rigorous testing is always recommended, regardless of application size, and you should consider asking others to test your program. Software bugs can be quite embarrassing.

Begin the testing process by using your program under normal conditions. Unfortunately, many developers stop here. Careful planning and design generally preclude mishaps when running the application under normal conditions. However, the causes of application crashes are special circumstances. You should actually try to crash your application by avoiding standard steps and procedures. For instance, what happens if the *USA_View* document is accidentally deleted by a user? What happens if a user creates several layouts with the same name?

Of course, these "what-ifs" can be escalated to an absurd level. It is up to you to decide how bullet-proof your application should be.

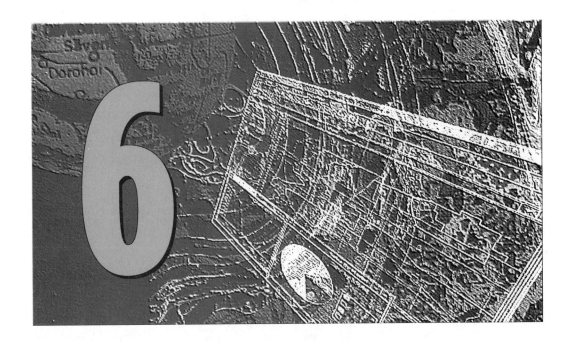

Programming View Documents

In this chapter a simplified crime analysis application will be created. The exercise includes creating a view document and adding graphical elements such as text.

Following an overview of the application, the Avenue scripts are developed. The customized interface required for the application is then reviewed. (See Chapter 4 for details on developing customized interfaces.)

Application Overview

The application to be developed consists of a base map and the following series of Avenue scripts.

❑ *CRM_Get_Locations* accepts a crime incident point from the view document through clicking on the mouse. The script then asks the user for the point label. A graphical point symbol and label text appear on the view document.

❑ *CRM_Analysis* draws a circle which incorporates all incident points. At least three points should be present. Lines are drawn from the center of this circle to each incident point.

The scripts described below will make the application "friendlier."

❑ *CRM_Scale1* and *CRM_Scale2* change the extent of the view display to one of two preset values. One view is at the neighborhood level, and the other is at the regional level.

❑ *CRM_New_Analysis* clears all graphical elements and resets all variables for a new set of incident points.

❐ *CRM_New_View* creates a new view document and loads a preset base map.

The customized interface consists of a tool button for the *CRM_ Get_Locations* script, and buttons for the remaining scripts. The button for *CRM_New_View* is placed in the project window user interface. The other control buttons are placed in the view document user interface, as shown in the following illustration.

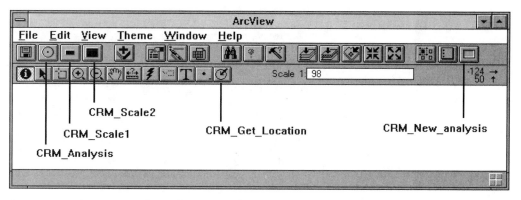

Customized user interface for view documents.

View Document

The first script to be developed is *CRM_New_View*. Create a new view document, set its properties and scale units, open its window, load a theme, and zoom to the extent of that theme.

CODE 06-01.AVE

```
' CRM_New_View
'
' Create a view document.
thisProject = av.GetProject
newView = View.Make
if (nil = newView) then
   MsgBox.Error
     ("Unable to create a new view", "CRM")
     exit
end
'
' Get a view name and set properties.
newViewName = MsgBox.Input
("Enter a new view name:", "CRM", "")
if (nil = newViewName) then
   MsgBox.Warning
     ("Program stopped by the user", "CRM")
     exit
end
newView.SetName ( newViewName )
newViewDpy = newView.GetDisplay
newViewDpy.SetUnits ( #UNITS_LINEAR_FEET )
'
' Display the document.
newViewWin = newView.GetWin
newViewWin.Open
'
' Load a theme.
areaMapName = SrcName.Make ("WARD_3.shp")
areaMapTheme = Theme.Make (areaMapName)
newView.AddTheme (areaMapTheme)
'
' Zoom to the view's extent.
newViewDpy.SetExtent(newView.ReturnExtent.Scale(1.1))
```

Two other scripts, *CRM_Scale1* and *CRM_Scale2*, are also developed in the initial phase. Because the only difference between the two is the scale value, only one script's development is shown here. You can easily develop the second one on your own.

06-02.AVE

```
' CRM_Scale1
'
' Set the extent of the current
' view to the neighborhood.
thisView = av.GetActiveDoc
thisDpy = thisView.GetDisplay
thisDpy.SetExtent(thisView.ReturnExtent.Scale(0.5))
```

Creating a New View

The scripts in your application typically act on an existing view prepared by the user. However, some applications could include predefined views created by scripts. Creating a view document is accomplished by sending a Make request to the View class. This request also adds the returned object to the project object. The following code segment illustrates this step.

✓ **TIP:** *Themes are added to the top of the table of contents (TOC) for a view. Therefore, if you add themes in the reverse order of the TOC, you do not have to reorder them.*

```
newView = View.Make
if (nil = newView) then
   MsgBox.Error
   ("Unable to create a new view","CRM")
   exit
end
```

Displaying a View Document

To open an existing document, you must have access to the document's window object. Once you have access, you can send an Open request to display it, as shown in the following code segment.

```
thisProject = av.GetProject
theView = thisProject.FindDoc ("View1")
theViewWin = theView.GetWin
if (theViewWin.IsOpen.Not) then
   theViewWin.Open
end
```

Setting View Properties

Avenue allows you to change the properties and parameters of a view. You can generally find a specific request, and send it to the view object to set a certain parameter. A list of requests used to set parameters and properties, such as name and projection, appears below.

```
aView.SetProjection (aProjection)
aView.SetTOCWidth (widthInPixels)
aView.SetName (aName)
aView.SetComments (commentString)
```

The following code segment sets the name and map units for the view.

```
newViewName = MsgBox.Input
("Enter a new view name:", "CRM", "")
if (nil = newViewName) then
  MsgBox.Warning ("Program stopped by the user", "CRM")
  exit
end
```

```
newView = View.Make
newView.SetName ( newViewName )
newView.GetDisplay.SetUnits ( #UNITS_LINEAR_FEET )
```

Setting the Display Extent

Avenue provides requests that change the extent of a displayed map. An application that uses multiple views to simultaneously display different parts of a map requires this type of request. The following code segment sets the extent to display all themes.

```
theView.GetDisplay.SetExtent (theView.ReturnExtent.Scale(1.1))
```

Other requests are available through the view's display object. In the following code segment, Avenue sets the display extent to the active themes.

```
theView = av.GetActiveDoc
themesList = theView.GetActiveThemes
aBox = Rect.MakeEmpty
for each oneTheme in themesList
   aBox = aBox.UnionWith (oneTheme.ReturnExtent)
end
theView.GetDisplay.SetExtent (aBox.Scale(1.1))
```

Graphical Elements

This section develops the *CRM_Get_Locations, CRM_ Analysis,* and *CRM_New_Analysis* scripts. The *CRM_ Get_Locations* script is associated with the apply property of a tool button and executes at the click of a mouse. The script places a point symbol at the coordinates where you click the mouse, and stores the points in a list tagged to the view so that *CRM_ Analysis* can access them. By tagging the list of points to the view document, the user can run *CRM_Analysis* at any time.

 06-03.AVE

```
' CRM_Get_Locations
'
' Get the display and graphic list for this view.
theView = Av.GetActiveDoc
theDpy = theView.GetDisplay
theGList = theView.GetGraphics
'
' Get the list object attached to the view,
' or create it if it does not exist.
pointList = theView.GetObjectTag
if (nil = pointList) then
   pointList = List.Make
   theView.SetObjectTag (pointList)
end
'
' Get a point.
aMousePoint = theDpy.GetMouseLoc
' Clone the mouse point to preserve its value.
aPoint = aMousePoint.Clone
aGPoint = GraphicShape.Make (aPoint)
'
' Get a label for this point.
aLabel = MsgBox.Input("Point label:","CRM","")
if (nil = aLabel) then
   MsgBox.Warning ("Point canceled", "CRM")
   exit
else
   aGLabel = GraphicText.Make (aLabel,aPoint)
end
'
' Add the point to the list and display.
theGList.Add (aGPoint)
theGList.Add (aGLabel)
pointList.Add (aPoint)
```

The *CRM_Analysis* script below uses the list of points attached to the view to determine a bounding rectangle for all points. The script then draws a circle through the four corners of the rectangle. Lines are also drawn from the center of the circle to each point on the list.

 06-04.AVE

```
' CRM_Analysis
'
theView = Av.GetActiveDoc
pointList = theView.GetObjectTag
if (nil = pointList) then
   MsgBox.Warning
   ("There are no points to analyze","CRM")
   exit
else
   theDpy = theView.GetDisplay
   theGList = theView.GetGraphics
end
'
' Must have at least three points to analyze.
if (pointList.Count < 3) then
   MsgBox.Error
   ("A minimum of three points is required","CRM")
   exit
end
' Establish the bounding rectangle
' by finding the minimum and maximum x and y values.
firstFlag = true
for each aPoint in pointList
   aX = aPoint.GetX
   aY = aPoint.GetY
   if (firstFlag) then
     minX = aX
     maxX = aX
     minY = aY
     maxY = aY
```

```
      firstFlag = False
    else
      minX = minX Min aX
      maxX = maxX Max aX
      minY = minY Min aY
      maxY = maxY Max aY
    end
end
'
' Verify that you can draw a circle through the min/max
' X and Y values.
if ((minX = maxX) OR (minY = maxY)) then
    MsgBox.Error
    ("Cannot analyze a linear distribution","CRM")
    exit
end
'
' Get the center and determine a radius.
midX = ((maxX - minX)/2) + minX
midY = ((maxY - minY)/2) + minY
cenPoint = Point.Make (midX,midY)
radLen = (((maxX-midX)^(2)) +
((maxY-midY)^(2))).Sqrt
'
' Draw a circle.
aCircle = Circle.Make (cenPoint,radLen)
if (nil = aCircle) then
    MsgBox.Error
    ("Unable to create circle", "CRM")
    exit
end
aGCircle = GraphicShape.Make (aCircle)
theGLIST.Add (aGCircle)
'
' Draw lines from the center to each point.
for each aPoint in pointList
    aLine = Line.Make (cenPoint, aPoint)
```

```
        aGLine = GraphicShape.Make (aLine)
        theGList.Add (aGLine)
end
```

The *CRM_New_Analysis* script removes existing points so that new points can be entered. The script selects and deletes all graphical elements and removes all points from the list attached to the view.

CODE 06-05.AVE

```
' CRM_New_Analysis
'
' Get the display and graphics list of this view.
theView = Av.GetActiveDoc
theDpy = theView.GetDisplay
theGList = theView.GetGraphics
'
' Attach a new and empty list for points.
pointList = List.Make
theView.SetObjectTag (pointList)
'
' Delete all graphical objects.
theGlist.SelectAll
theGlist.ClearSelected
```

Graphics List

You can draw point symbols, lines, text, and other graphical elements in a view document on top of any displayed theme. ArcView holds these elements in a list associated with the view's display. Avenue can access and manipulate the associated list. Once you have accessed the list, you can add or delete elements, or change the size, location, and orientation of elements. The following request accesses a view's graphical list.

```
aGList = theView.GetGraphics
```

Appearing below are samples of requests that can be made to a graphics list.

```
aGList.Add (aGraphicElement)
howMany = aGList.Count
aNewGList = aGList.GetSelected
aGList.SelectAll
aGList.UnselectAll
aGList.ClearSelected
```

Drawing Graphical Elements

Four steps are required to draw a graphical element on an existing view.

1. Retrieve the associated graphics list.

2. Create the object to draw (e.g., a line or circle).

3. Convert the object to a graphical element.

4. Add the element to the graphics list.

The code segment below shows how lines are drawn from a central point to various points stored in a list.

```
theGlist = theView.GetGraphics
' pointList holds objects of the Point class.
for each aPoint in the pointList
    aLine = Line.Make (cenPoint, aPoint)
    aGLine = GraphicShape.Make (aLine)
    theGList.Add (aGLine)
end
```

Text works in a similar fashion, as shown in the following code segment.

```
theGlist = theView.GetGraphics
aLabel = MsgBox.Input ("Point label:", "CRM", "")
aGLabel = GraphicText.Make (aLabel,aPoint)
theGList.Add (aGLabel)
```

Shapes that can be used to create graphical elements are listed below.

```
aCircle = Circle.Make (centerPoint, radiusLength)
aGCircle = GraphicShape.Make (aCircle)
aLine = Line.Make (startPoint, endPoint)
aGLine = GraphicShape.Make (aLine)
anOval = Oval.Make (originPoint, extentPoint)
aGOval = GraphicShape.Make (anOval)
aPoint = Point.Make (x, y)
aGPoint = GraphicShape.Make (aPoint)
aPolygon = Polygon.Make (aListOfVertices)
aGPolygon = GraphicShap.Make (aPolygon)
aRectangle = Rect.Make (originPoint, extentPoint)
aGRectangle = GraphicShape.Make (aRectangle)
aGText = GraphicText.Make (aTextString, lowerLeftPoint)
aGText.SetBounds (aRectangle)
```

The SetBounds request defines the extent of a graphical element. When used with text, the request defines paragraph margins.

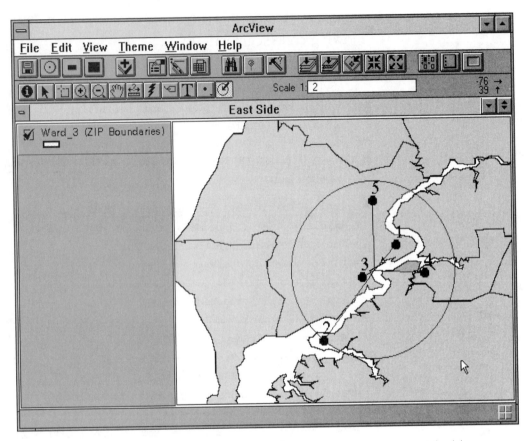

A simplified crime analysis application. (The coverage was provided by Geographic Data Technology, Inc.)

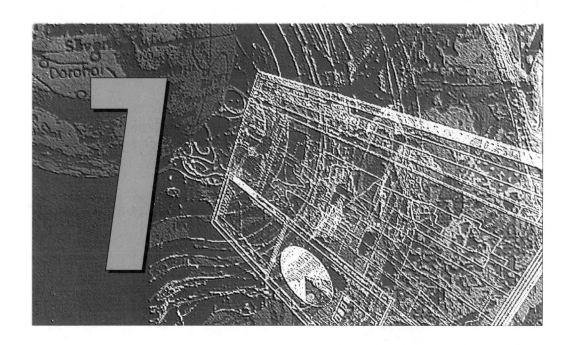

Programming Themes

The ability to access and manipulate objects through Avenue scripts adds powerful capabilities to any ArcView application. This chapter explores how to add themes, change theme properties, and select theme features.

Adding and Displaying Themes

Applications containing predetermined sets of themes are common. Your application should automatically add the theme and set its properties. This section demonstrates how to add a theme to a view, duplicate it with different properties, and work with the theme legend. A script named *Load_Theme* is developed to demonstrate these features. Results of running the script are shown in the following figure.

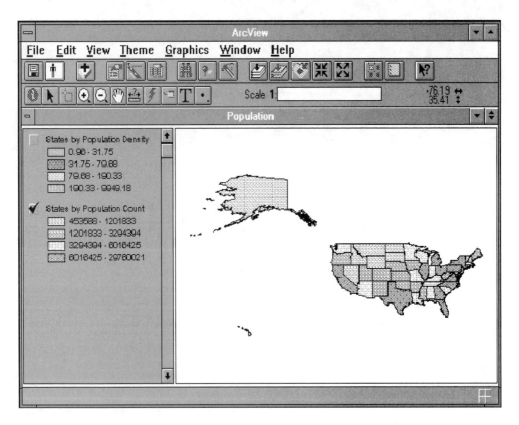

Adding and displaying themes.

Script features are explained within the code segment below. You may have to replace the file path with your path data.

 07-01.AVE

```
' Load_Theme
'
' This script is executed from a customized
' button on the view document.
'
theView = Av.GetActiveDoc
'
' Establish a data source and load it into the view.
aSrcName = SrcName.Make
("d:/progs/esri/arcview/ver30/esridata/usa/states.shp")
aTheme = Theme.Make (aSrcName)
theView.AddTheme (aTheme)
'
' Duplicate the new theme, then
' change its name.
aTheme.SetActive (True)
theView.CopyThemes
theView.Paste
firstTheme = theView.GetThemes.Get(0)
secondTheme = theView.GetThemes.Get(1)
firstTheme.SetName ("States by Population Density")
secondTheme.SetName ("States by Population Count")
'
' Load the table with population data.
popVTab = VTab.Make
("f:/progs/esri/arcview/ver30"+
"/esridata/usa/tables/stdemog.dbf"
.AsFileName,
False, False)
'
' Join the themes with the population table.
popJoinField = popVTab.FindField
("State_name")
```

```
theDensityVTab = firstTheme.GetFTab
theCountVTab = secondTheme.GetFTab
for each oneVTab in {theDensityVTab,theCountVTab}
   theJoinField = oneVTab.FindField
   ("State_name")
   oneVTab.Join (theJoinField, popVTab, popJoinField)
end
'
' Set each theme's legend.
' Use quantile for the first legend and
' equal intervals for the second.
firstLegend = firstTheme.GetLegend
firstLegend.SetLegendType
(#LEGEND_TYPE_COLOR)
firstLegend.Quantile (firstTheme, "Pop90_sqmi", 4)
secondLegend = secondTheme.GetLegend
secondLegend.SetLegendType
(#LEGEND_TYPE_COLOR)
secondLegend.Quantile (secondTheme, "Pop1990", 4)
'
' Display the population count theme
' and make sure the legends are visible.
firstTheme.SetVisible (False)
firstTheme.SetLegendVisible (True)
secondTheme.SetVisible (True)
secondTheme.SetLegendVisible (True)
```

Data Source

Themes are created from data sources. ArcView supports a variety of data sources, including raster image files, ArcStorm, or dynamic segmentation events. ArcView encapsulates these diverse data sources into objects of the SrcName class. A SrcName object identifies the data source used in creating a theme. The SrcName class has specialized subclasses to accommodate more complex data sources. ArcView data sources and identifying classes are listed below.

- ❏ Image file, SrcName

- ❏ Shape file, SrcName

- ❏ Coverage, SrcName

- ❏ Library layer, SrcName

- ❏ ArcStorm, SrcName

- ❏ X-Y event file, XYName

- ❏ Geocoding feature source, GeoName

- ❏ Dynamic segmentation source, DynName

A SrcName object has four attributes: file name, name, sub-name, and data source name. The file name attribute references the disk location of the data source, and is nil for a library layer. The name attribute is the name of the data source which becomes the name of the theme created from the SrcName. The sub-name attribute constitutes the feature class, that is, region, route, annotation, polygon, arc, or point. Finally, the data source is the name of the coverage, file, or library layer that contains the data. For coverages or files, the data source includes the full path to the coverage or file. In the case of a library layer, the data source has the following form: *LibraryName.LayerName*.

You can create a SrcName object with the Make request. The Make request requires a source string as its parameter, which varies for different data source types. The following code segment shows how to create a SrcName object for each source type.

```
' Image file.
aSrcName = SrcName.Make ("c:\images\myimage.bil")
'

' Shape file.
aSrcName = SrcName.Make ("c:\shapes\myshape.shp")
'

' Polygon coverage.
aSrcNamne = SrcName.Make ("c:\covrgs\usa polygon")
'

' ARC/INFO librarian.
aSrcName = SrcName.Make ("library.layer polygon")
'

' ArcStorm.
aSrcName = SrcName.Make
("librarian.library.layer polygon")
'

' A region.
aSrcName = SrcName.Make
("c:\avdata\usa region.state")
```

Once a SrcName object is created you can add it to a view as a theme. The following statements show how to add a theme from the SrcName object.

```
aTheme = Theme.Make (aSrcName)
aView.AddTheme (aTheme)
```

The SrcName class also provides requests that retrieve information about an existing source name. If you are uncertain about the required properties, you can manually load your theme. Use the following Avenue script to learn more about the properties of the theme's source name.

 07-02.AVE

```
' Get_SourceName_Info
'
' This script is executed from a customized
' button in a view document. It assumes
' that your theme is the first one in the
' view's table of contents.
theView = Av.GetActiveDoc
theTheme = theView.GetThemes.Get(0)
theSrcName = theTheme.GetSrcName
'
' Get source name's properties.
theName = theSrcName.GetName
theFile = theSrcName.GetFileName
theDataSource = theSrcName.GetDataSource
theSubName = theSrcName.GetSubName
'
' Place these properties in a list to display.
aList = List.Make
aList.Add ("Name:"++theName)
aList.Add ("File:"++theFile.GetFullName)
aList.Add ("Source:"++theDataSource)
aList.Add ("Sub:"++theSubName)
MsgBox.ListAsString
(aList,"Source name properties:","")
```

Legends

There are two ways to establish legends. One way is to manually create a legend in ArcView and store it as a legend file for later retrieval and use in Avenue. The other way is to access a theme's legend and modify it through object requests.

The following code shows how to load a legend file. This approach is useful in cases where the legend does not change every time the application is used, or where a basic legend is the starting point for creating a more complex and dynamic version.

 07-03.AVE

```
' Load_Legend
'
' This script is executed from a customized
' button in a view document, and assumes
' that the legend applies to the first theme
' on the view's table of contents.
theView = Av.GetActiveDoc
theTheme = theView.GetThemes.Get(0)
theLegend = theTheme.GetLegend
aLegendFile = "c:/temp/myleg.avl".AsFileName
theLegend.Load (aLegendFile,#LEGEND_LOADTYPE_ALL)
' Apply the legend changes.
theTheme.InvalidateLegend
```

Copying

A theme is a pointer to a data source; therefore, copying a theme does not copy the data but simply generates new pointers. Theme copying provides a powerful feature to view diverse aspects of a single data source.

Themes can be independently queried to show a particular aspect of the data source. For example, a point coverage that contains insurance company agents and the years their offices were opened can hold several themes displaying offices opened in a specific year or years.

CopyThemes and Paste requests are sent to a view object to copy the active themes to a clipboard, and then paste them to the view at the beginning of the view's table of contents. In the following code segment, all themes are copied from one view to another.

```
' Make all themes active.
themesList = theView.GetThemes
for each aTheme in themesList
   aTheme.SetActive (True)
end

theView.Copythemes
anotherView.Paste
```

The CutThemes request removes the active themes from the view and places them on the clipboard.

Using Queries

Avenue can set and evaluate a query expression for a theme. Applications requiring dynamically changing queries can use this feature. In the following script, Avenue sets a query that would select all states with equal or greater population than a preselected state.

CODE 07-04.AVE

```
' Set_Query
'
' This script is executed from a button on
' the view's user interface. A state
' must be selected beforehand.
'
' Get the basic information.
theView = Av.GetActiveDoc
theTheme = theView.FindTheme ("States.shp")
theDpy = theView.GetDisplay
```

```
theVTab = theTheme.GetFTab
theBitMap = theVTab.GetSelection
'
' Retrieve the population of the
' selected state.
if (theBitMap.Count <> 1) then
   MsgBox.Warning ("Select one state","Set_Query")
   exit
end

selState = theBitMap.GetNextSet (-1)
popField = theVTab.FindField ("Pop1990")
selPop = theVtab.ReturnValueString (popField, selState)
'
' Set a query for states of equal or
' greater population. Place field names
' in square brackets [ ].
aQStr = "[Pop1990] >="++selPop
' Apply the query to the theme.
theVTab.Query (aQStr, theBitMap, #VTAB_SELTYPE_NEW)
theVTab.SetSelection (theBitMap)
```

Selecting Features

ArcView's default user interface for view documents contains a tool button for selecting features. In Avenue, features can be selected by accepting a point, a line, a rectangle (box), or a polygon from the user.

Point Selection

SelectByPoint is the request sent to a theme to select a spatial feature. In the following code segment, a feature is selected from a location identified by pointing the mouse. The feature is added to the list of selections.

07-05.AVE

```
' Point_Selection
'
' This script is executed from a tool button
' of a view document. It selects features from
' the first theme on the table of contents.
theView = Av.GetActiveDoc
'
' The view's display object is required to get a
' mouse location.
theDpy = theView.GetDisplay
'
' Get the first theme.
theTheme = theView.GetThemes.Get(0)
'
' Make the selection.
theTheme.SelectByPoint
(theDpy.GetMouseLoc, #VTAB_SELTYPE_OR)
```

The second parameter in the SelectByPoint request determines whether the selection set is new or added to current selections. Valid values for this parameter follow.

```
' Create a new selection set.
#VTAB_SELTYPE_NEW

' Select from the current selection set.
#VTAB_SELTYPE_AND

' Add to the current selection set.
#VTAB_SELTYPE_OR

' If the selected feature is not currently
' selected, then add it to the current
' selection set. Otherwise, remove it from
' the current selection set.
#VTAB_SELTYPE_XOR
```

The first parameter for the SelectByPoint request is a point object. GetMouseLoc returns a point which is the location of a mouse click.

Line Selection

Selecting features with a line is similar to point selection.

```
theTheme.SelectByLine (theDpy.ReturnUserLine, #VTAB_SELTYPE_NEW)
```

Box Selection

The SelectByRect request selects features either inside a rectangle or touched by the rectangle's sides. You can supply the rectangle by constructing it, or the user can select it. In the following script, the user selects a rectangular area.

```
' Box_Selection
'
' This script is executed from a tool button
' of a view document. It selects features from
' the first theme on the table of contents.
theView = Av.GetActiveDoc
'
' Get the first theme.
theTheme = theView.GetThemes.Get(0)
'
' Ask user for a rectangle.
aBox = theView.ReturnUserRect
'
' Make the selection.
theTheme.SelectByRect (aBox, #VTAB_SELTYPE_NEW)
```

Polygon Selection

Selecting features with a polygon is similar to the box selection. The difference is in the request. The following statement selects features using a polygon chosen by the user.

```
theTheme.SelectByPolygon (theView.ReturnUserPoly, #VTAB_SELTYPE_NEW)
```

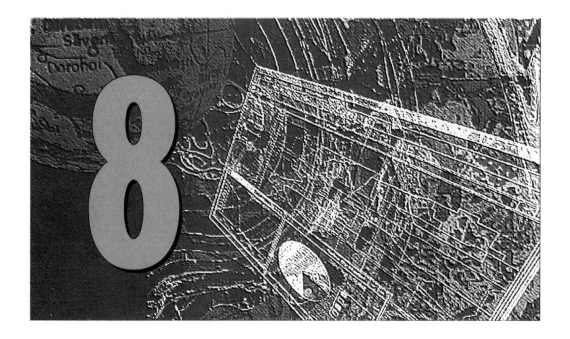

Programming Table Documents

Because ArcView maintains attribute data in Table documents, attribute data must be accessed and manipulated through these documents. In this chapter, an application is developed that assigns the polygons of a coverage to territories or districts. This type of application is useful for redistricting, assigning sales territories, and similar projects. In the process, you will learn how to create tables, retrieve or set attribute values, and join tables.

Assigning Polygons to Territories

Avenue scripts in this chapter use the *states.shp* shape file bundled with the ArcView software. The shape file is located in ArcView's *usa* directory. All scripts assume that the *States.shp* is the first theme in a view.

The scripts described below are developed in subsequent code segments. Script results are shown in the figure following the code.

Create_Att_Table creates a new attribute table with two fields, one for the join operation and the other to hold a territory number. Because this script is used only once for each new coverage, execution from a menu item is more appropriate than from a button.

Load_Att_Table imports the newly created attribute table and joins it with the theme's table. This script is also executed from a menu item because it is needed only once for each new coverage.

Assign_Territory sets the territory field number of the selected polygons to a user-provided value. The user can pick any of the selection tools, such as point, circle, or polygon, to select appropriate polygons. The script deselects all polygons after setting the territory number.

CODE 08-01.AVE

```
' Create_Att_Table
'
' This script creates a new table and copies
' values of a field to be joined later.
'
' Get a file name to create.
aFileName = FileDialog.Put("terr.dbf".AsFileName,
"*.dbf","Create Attribute Table")
if (nil = aFileName) then
   exit
end
'
' Create a dBase table.
attVTab = VTab.MakeNew (aFileName,dBase)
attVTab.SetEditable (True)
fipsField = Field.Make
("State_fips", #FIELD_CHAR, 2, 0)
terrField = Field.Make
("Terr_num", #FIELD_CHAR, 4, 0)
attVTab.AddFields ({fipsField, terrField})
'
' Get State_fips values from the theme's table
' and write to the new table.
thisView = Av.GetActiveDoc
theTheme = thisView.GetThemes.Get(0)
theVTab = theTheme.GetFTab
stateField = theVTab.FindField("State_fips")
if (nil = stateField) then
   MsgBox.Error ("Unable to find the fips field",
   "")
   exit
end
'
' The record count is required for a loop.
recordCount = theVTab.GetNumRecords
if (recordCount = 0) then
   MsgBox.Error ("There are no records to copy",
```

```
      "")
      exit
end
      '

' Read and write each record.
for each index in theVTab
   fieldValue =
   theVtab.ReturnValueString (stateField,index)
   if (nil = fieldValue) then
      continue
   end
   ' Add territory number zero to each
   ' record as default.
   rec = attVTab.AddRecord
   attVTab.SetValue (fipsField, rec, fieldValue)
   attVTab.SetValue (terrField, rec, "0")
end
attVTab.Flush
```

08-02.AVE

```
' Load_Att_Table
'
' This script imports the attribute table and
' joins it with the theme's table.
'
' Get the theme's table.
thisView = Av.GetActiveDoc
theTheme = thisView.GetThemes.Get(0)
theVTab = theTheme.GetFtab
theJoinField = theVtab.FindField("State_fips")
if (nil = theJoinField) then
   MsgBox.Error
   ("Theme's table does not have the join field",
   "")
   exit
end
      '
```

```
' Load the attribute table.
attTableFile = FileDialog.Show
("*.dbf","dBase Files",
"Load Attribute Table")
if (nil = attTableFile) then
   exit
end
'
' Verify that a table can be created
' from the given file name.
isOK = VTab.CanMake (attTableFile)
if (isOK.Not) then
   MsgBox.Error ("Invalid file","")
   exit
end
'
' Load the table.
forWrite = True
skipFirst = False
attVTab = VTab.Make (attTableFile, forWrite, skipFirst)
attTable = Table.Make (attVTab)
if (nil = attTable) then
   MsgBox.Error ("Unable to load attribute table","")
   exit
end
attTable.SetName ("Territory")
attJoinField = attVTab.FindField ("State_fips")
if (nil = attJoinField) then
   MsgBox.Error
   ("Attribute table does not have the join field","")
   exit
end
'
' Join the tables and display the result.
theVTab.Join (theJoinField, attVTab, attJoinField)
theTheme.EditTable
```

08-03.AVE

```
' Assign_Territory
'
' This script will set the selected records
' to a specified territory number.
thisView = Av.GetActiveDoc
theTheme = thisView.GetThemes.Get(0)
theVTab = theTheme.GetFTab
theVTab.SetEditable (True)
isOK = theVTab.IsEditable
if (isOK.Not) then
   MsgBox.Error ("Cannot edit the table", "")
   exit
end
'
' Verify that at least one feature is selected.
theBitMap = theVTab.GetSelection
selectedCount = theBitMap.Count
if (selectedCount = 0) then
   MsgBox.Warning ("Select areas first", "")
   exit
end
'
' Get a territory number.
terrNum = MsgBox.Input ("Enter a territory number: ",
"","0")
if (nil = terrNum) then
   exit
end
'
' Get the field to edit.
attVTab = av.GetProject.FindDoc("Territory").GetVTab
attVTab.SetEditable (True)
theField = attVTab.FindField("Terr_num")
theField.SetEditable(True)
'
' Set the value of the selected records.
for each index in theVTab.GetSelection
   attVTab.SetValueString(theField, index, terrNum)
end
```

```
'
' Clear the selection.
theBitMap.ClearAll
theVtab.UpdateSelection
```

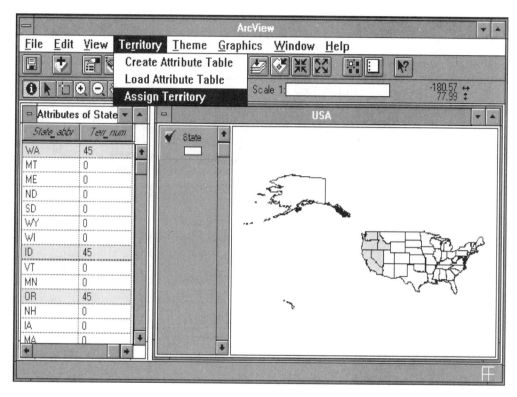

Assigning polygons to a territory.

Table Document

Tables provide the means to manipulate the data associated with spatial elements. You will need to access such data in most of your applications. Tables are document objects that serve as an interface to the underlying database.

Upon opening a Table document, a VTab is displayed in a format resembling a spreadsheet. VTab objects are virtual tables composed of tabular data stored on disk outside ArcView.

Creating a Table

You can create a Table document from existing disk files or from new disk files. If you are creating a new table in a new disk file, your choices are comma-delimited text, dBase, or INFO files. In the case of creating Table documents from existing files, your choices include an SQL table, dBase files, INFO files, and comma-delimited text files.

To create a new comma-delimited text file, use the FileDialog class to establish a file name. Create the file and write the field names on the first line as shown in the following code segment.

```
' Get a file name to create.
aFileName = FileDialog.Put("terr.txt".AsFileName,
"*.txt","Create Attribute Table")
if (nil = aFileName) then
   exit
end


'
' Create a delimited text file.
aLineFile = LineFile.Make(aFileName, #FILE_PERM_WRITE)
'
' Write field names as the first line.
aLineFile.WriteElt("State_abbr,Terr_num")
```

You can continue writing data into the new file. However, once the file is created it can be treated as an existing file. The code segment below creates a new table document by loading a disk file into the ArcView project.

```
newVTab = VTAB.Make (aFileName, forWrite, skipFirst)
newTable = Table.Make (newVTab)
```

Displaying a Table

The process of opening or closing a Table document can be automated through Avenue. If the table is associated with a feature theme, you can use the EditTable request to open the theme's table as shown below.

```
aFeatureTheme.EditTable
```

A more direct method to control document display is through its window. As demonstrated in the following code segment, once you have a table's DocWin object, you can then open, close, resize, and move the table, as well as carry out other operations.

```
' attTable is a table document.
' Get its document window object.
attDocWin = attTable.GetWin
'
' Open the object if it is closed.
if ( attDocWin.IsOpen.Not ) then
   attDocWin.Open
end
'
' Minimize the table.
attDocWin.Minimize
'
' Restore the table from its icon form.
```

```
attDocWin.Restore
'

' Make the table the active window.
attDocWin.Activate
'

' Move the document around.
attDocWin.MoveTo ( newX, newY)
'

' Close the document.
attDocWin.Close
```

Setting Table Properties

Table properties include name, creator, and comments. Similar to other document properties, table properties are generally used for documentation. At the very least, you should set the name property because this is always visible. Setting the creator and comments properties is recommended.

The following code lines show the Avenue requests for setting table properties.

```
' newTable is a Table object.
' Set its properties.
'

newTable.SetName ("District Detail")
newTable.SetCreator ("A. H. Razavi")
newTable.SetComments
("Store characteristics of"++
"each district in this table")
```

As demonstrated below, you can access the properties of an existing table through Get requests.

```
' Table name is a string object.
aName = aTable.GetName
```

```
whoCreated = aTable.GetCreator
whenCreated = aTable.GetCreationDate
tableComments = aTable.GetComments
```

Joining Tables

Spatial analyses often require information from more than one table. Multiple tables to be used simultaneously must be joined. The most frequent join operation involves a theme's table and a non-spatial attribute data file. Avenue can automate the join process for the user. In the following code segment, the join fields from two tables are identified, and then the two tables are joined through their respective VTabs.

```
' Get the theme's table information.
thisView = Av.GetActiveDoc
theTheme = thisView.GetThemes.Get(0)
theVTab = theTheme.GetFtab
theJoinField = theVtab.FindField("State_fips")
'
' Get the attribute table information.
attVTab = attTable.GetVTab
attJoinField = attVTab.FindField ("State_fips")
'
' Join the tables and display the result.
theVTab.Join (theJoinField, attVTab, attJoinField)
```

As presented in the following code line, the Join request for a VTab requires three parameters. The parameters are described below.

```
aToVtab.Join (aToField, aFromVtab, aFromField)
```

❐ *aToField* is the common field from the destination table which receives the result of the join operation (*theJoinField* in the preceding example).

❏   *aFromVTab* is the source table for the join operation (*attVTab* above).

❏   *aFromField* is the common field from the source table (*attJoinField* in the example).

To join more than two tables, first join two tables, then join the third table to the result, and so on. A join operation assumes that the relationship between the two tables is one-to-one. However, the relationship may also be one-to-many, where each record of the source table matches one or more records of the destination table.

If the relationship is many-to-one, that is, when one or more matching records in the source table exist(s) for each record in the destination table, you must use the link instead of the join operation. A link operation may also be applied when the relation is one-to-one or one-to-many. Link does not provide a virtual table comprised of the source and destination tables. However, the link operation does return the linked information for the Select or Identification operations. The syntax is similar to a join, as shown below.

```
aToVTab.Link (aToField, aFromVTab, aFromField)
```

Spatial joins are also performed in the preceding fashion. Use the Join request on two objects of the FTab class. Instances of the FTab class are specialized VTab objects that model attributes of feature themes. They can be accessed by applying a GetFTab request to a feature theme. Next, use the Shape field from both tables in the join operation, as shown below.

```
aToFTab.Join (toShapeField, aFromFTab, fromShapeField)
```

Sorting and Summarizing Records

Sorting a table can improve readability, and may be required in your application. An example would be displaying states in ascending order of population size. The Sort request accepts the sort field and Boolean parameters. If the Boolean parameter is True, the sort is arranged in descending order; if False, the sort will be arranged in ascending order.

```
myTable.Sort (sortField, isItDescending)
```

ArcView creates a new file when you summarize a table. The new file maintains the result of the summary. The syntax for summarizing is shown below.

```
anFTab.Summarize (newFileName, newFileType,
groupField, summaryFieldList, summaryOperationList)
```

The new file type can be dBase, INFO, shape, or DText for delimited text file. The summary is grouped by the unique values in the groupField. The fields provided in the summaryFieldList match the order of the summary operation in summaryOperationList.

The summary operations are identified by the following enumerations.

❏ **#VTAB_SUMMARY_AVG**: Average of a group.

❏ **#VTAB_SUMMARY_COUNT**: Count of non-null values in a group.

❏ **#VTAB_SUMMARY_FIRST**: First value in a group.

- ❑ **#VTAB_SUMMARY_LAST**: Last value in a group.

- ❑ **#VTAB_SUMMARY_MAX**: Maximum value in a group.

- ❑ **#VTAB_SUMMARY_MIN**: Minimum value in a group.

- ❑ **#VTAB_SUMMARY_STDEV**: Standard deviation of a group.

- ❑ **#VTAB_SUMMARY_VAR**: Variance in a group.

You can use the shape field as a summary field to merge shapes. The following script uses the *States* shape file to create a new shape file for the entire USA.

08-04.AVE

```
' Access the states feature table.
theView = av.GetProject.FindDoc ("View1")
statesTheme = theView.FindTheme ("States.shp")
statesFTab = statesTheme.GetFTab
statesFTab.SetEditable (true)
shapeField = statesFTab.FindField ("Shape")
'
' Create a new field that will have the
' same value for all states.
groupField = Field.Make ("Group", #FIELD_SHORT, 1, 0)
statesFTab.AddFields ({groupField})
for each rx in statesFTab
    statesFTab.SetValue (groupField, rx, 0)
end
'
' Summarize.
statesFTab.Summarize ("usa".AsFileName, shape,
groupField,{shapeField},{#VTAB_SUMMARY_AVG})
```

```
' Load the new shape file.
usaSrcName = SrcName.Make("usa.shp")
usaTheme = Theme.Make (usaSrcName)
theView.AddTheme (usaTheme)
```

Note that the summary operation is #VTAB_SUMMARY_ AVG instead of merge. There is no enumeration element for merging. You can select any of the operations, but ArcView only merges when it comes to the shape file regardless of the requested operation.

Printing Tables

Generating reports requires the ability to query the table, select fields, and format the results. ArcView allows the user to apply queries to tables, and through Avenue you can output selected fields. Formatting the report, however, is not possible. Consider exporting the entire table or selected records to an external file, so that third party software could be used to generate reports.

To print a table into an external file, use the Export request as demonstrated in the following script.

CODE   08-05.AVE

```
' Export_Table
'
' This script exports the themes table.
thisView = Av.GetActiveDoc
theTheme = thisView.GetThemes.Get(0)
theVTab = theTheme.GetFTab
if (nil = theVTab) then
    MsgBox.Error("Unable to open theme's table",
    "")
exit
end
'
```

```
' Get a file name.
aFileName = FileDialog.Put("terr.dbf".AsFileName,
"*.dbf","Create External Table")
if (nil = aFileName) then
   exit
end
'
' Export the table to a dBase file.
theVTab.Export (aFileName, dBase, False)
```

As seen in the following code lines, the Export request requires three parameters. The parameters are described below.

```
aVTab.Export (aFileName, aTableClass,
selectedRecordsOnly)
```

❑   *aFileName* is the name of the destination file for the export operation (*aFileName* in the preceding example).

❑   *aTableClass* can be a dBase, INFO, or DText value (*dBase* above).

❑   *selectedRecordsOnly* is a Boolean object that determines if the entire table is exported, or only selected records (*False* in the preceding example). A False value exports the entire table.

Accessing and Editing Databases

Tabular databases outside ArcView can be viewed and sometimes edited through table documents. ArcView tables do not hold actual data, but rather manage a view of a tabular data source. Tables are also dynamic, meaning that changes to the data source are reflected in the ArcView table. In addition, edits to the table document directly affect the source.

Chapter 8 reviewed the importance of attribute data and how to use such data in your applications. This chapter explores how to access data from a variety of sources. ArcView can use data sources such as dBase, INFO, and delimited text files. It can also connect to SQL databases and run SQL queries. Supported SQL databases include AS400, Informix, Ingres, Oracle, and Sybase. The last section of this chapter shows how to access ArcStorm, the ARC/INFO database.

Creating File Based Database Objects

The purpose of a table document in ArcView is to display a VTab. The VTab object, also known as a "virtual table," manages the view of the tabular data source. VTabs may be joined or linked to form new views of the data. These operations were discussed in Chapter 8.

To create a database object, you must first create a VTab representing the data. A VTab object can be created based on an existing data source, or a new data source can be simultaneously created. The Avenue script below shows how to create a VTab object from an existing data source or sources.

CODE   09-01.AVE

```
' Ask user to select existing
' data source(s) from
' dBase, or delimited text. Add
' each data source to the project.
fileLabels = {"dBase","Delimited Text", "All Files"}
filePatterns = {"*.dbf","*.txt", "*.*"}
fileNameList = FileDialog.ReturnFiles
(filePatterns, fileLabels,
"Select Data Sources", 0)
for each aFileName in fileNameList
```

```
    if ( VTab.CanMake (aFileName) ) then
      ' Create a VTab in read only mode.
     aVTab = VTab.Make (aFileName, False, False)
     if (aVTab.HasError) then
       MsgBox.Error
       ("Unable to create a database object from"++
       aFileName.AsString, "")
     else
       ' Create a table from the VTab.
       aTable = Table.Make (aVTab)
       aTable.SetName (aFileName.GetBaseName)
     end
    end
end
```

In the preceding script, the Make request to the VTab class requires the three parameters listed below.

```
myVTab = VTab.Make (aFileName,
forWrite,
skipFirstRecord)
```

If set to True, the second parameter creates the database as an editable object. The value of this parameter is returned by the CanEdit request. However, setting this parameter does not guarantee that a VTab can be edited. The parameter acts as a security measure to allow editing. You must first ask to begin editing by using the SetEditable request. The IsEditable request returns True if the VTab object can be edited. Text files cannot be edited. Also be aware of file permissions when accessing tables on the network. You must have write-permission to the file and directory before editing. These requests are presented below.

```
' allowedToEdit and canModify are
' Boolean objects.
allowedToEdit =
```

```
myVTab.CanEdit
myVTab.SetEditable (True)
canModify = myVTab.IsEditable
```

If set to True, the third parameter in the Make request causes VTab to skip the first record. This feature is often useful with text files that hold field names in the first line of the file.

✓ **TIP:** *You can also use the third parameter of the Make request for a VTab to skip the universe-polygon in ARC/INFO coverages.*

Two other new requests, CanMake and HasError, have also been used in the preceding script. The Can-Make request to the VTab class returns True if a VTab can be created from a file name; otherwise, it returns False. The HasError request to a VTab object returns True if there is a problem in creating the VTab object; otherwise, it returns False.

A new data source file can be created by using Ave-nue. New files are often created to export existing data. The following Avenue script exports selected records in a table document.

CODE 09-02.AVE

```
' Ask user to select an export format
' from INFO, dBase, or delimited text.
' Set a new file name, create table,
' and export the selected records.
' This script is associated with the
' click property of Table user interface.
theTable = av.GetActiveDoc
fileFormat = MsgBox.ChoiceAsString
({"INFO","dBase","Delimited Text"},
"Select an export format:", "")
```

```
if (fileFormat = "INFO") then
   fileClass = INFO
   filePattern = "arcdr9"
   fileExtension = ""
elseif (fileformat = "dBase") then
   fileClass = dBase
   filePattern = "*.dbf"
   fileExtension = "dbf"
elseif (fileformat = "Delimited Text") then
   fileClass = DText
   filePattern = "*.txt"
   fileExtension = "txt"
else ' User clicked on Cancel button.
   exit
end
' Get a new file name based on the selected format.
newFileName = FileDialog.Put
(("new."+fileExtension).AsFileName,
filePattern, "Export to a New File")
if (nil = newFileName) then
   ' User clicked on Cancel button.
   exit
end
theVTab = theTable.GetVTab
' Export selected records.
theVTab.Export (newFileName, fileClass, True)
```

The Export request to a VTab object requires three parameters. The first parameter is a new file name. The second determines the type of file, that is, INFO, dBase, or delimited text. The third parameter is a Boolean object. If the Boolean object is set to True, only the selected records are exported; if set to False, all records are exported. The structure of the new database file is the same as the Vtab's structure.

You can create a new database file and then directly build its structure by adding fields. In the following script, a dBase file is created to maintain county populations.

09-03.AVE

```
' The new dBase file has four fields.
' Create the field objects, create the dBase
' file, and then add the fields to the file.
field1 = Field.Make
("COUNTY", #FIELD_CHAR, 20, 0)
' Total population field.
field2 = Field.Make
("POP_TOTAL", #FIELD_LONG, 8, 0)
' Population age 18 years and over.
field3 = Field.Make
("POP_18", #FIELD_LONG, 8, 0)
' Percent of 18+ population.
field4 = Field.Make
("PCT_18", #FIELD_DECIMAL, 6, 2)
' Move all fields into a field list.
fieldList = {field1, field2, field3, field4}
' Create a new data source file and object.
newVTab = VTab.MakeNew ("c:/temp/pop.dbf".AsFileName,dBase)
if (newVTab.HasError) then
   exit
end
' Add the fields to the new dBase file.
if (newVTab.CanAddFields) then
   newVTab. AddFields (fieldList)
end
```

The Make request to the Field class requires four parameters. The first and second parameters are the field name and the field type enumeration. The third and fourth parameters are the width and precision of the field. For example, in the preceding script,

PCT_18 is a decimal field of six digits with two decimal places.

The MakeNew request to the VTab class requires two parameters: a file name, and a file class. File classes include INFO, dBase and DText. The file name is overwritten if it already exists.

Reading Records and Fields

Once a VTab object is created from a tabular data source, the VTab can access the entire database. A table document displays the records accessed by the VTab. The Vtab's view of data records is limited only if a definition query is associated with the VTab.

In this section, the discussion includes how to access field values for use inside an application. In the following code segment, the difference between states with the highest and lowest populations is computed.

09-04.AVE

```
' To retrieve a field value, the
' field object is required.
theTable = av.GetProject.FindDoc("stdemog.dbf")
theVTab = theTable.GetVTab
stateField = theVTab.FindField ("State_name")
popField = theVTab.FindField ("Pop1990")
' Get the highest and lowest values by sorting.
' A descending sort places the highest value in
' row 1.
' You must open the table before sorting.
theTable.GetWin.Open
theTable.Sort (popField, True)
recordNumber = theTable.ConvertRowToRecord (0)
mostState = theVtab.ReturnValueString
(stateField, recordNumber)
mostPopulation = theVTab.ReturnValueNumber
```

```
(popField, recordNumber)
' The lowest value is in the last row.
recordNumber = theTable.ConvertRowToRecord
((theVTab.GetNumRecords)-1)
leastState = theVtab.ReturnValueString
(stateField, recordNumber)
leastPopulation = theVTab.ReturnValueNumber
(popField, recordNumber)
' Report the difference.
MsgBox.Info (mostState++"has"++
(mostPopulation-leastPopulation).SetFormat("d").AsString++
"more people than"++leastState, "")
```

When a table sorts or promotes VTab records, record order does not physically change. Instead, changes in record order are displayed. Thus, there is a difference between table document and VTab rows. In the preceding code segment, the table was sorted in such a way that the record with the highest population was placed in the first row. However, the sort did not change the record number in VTab. The ConvertRow-ToRecord request was used for this purpose. When the row of the least populated state was converted to record number, GetNumRecords-1 was used as the number for the last row in the table because the rows start at zero. The GetNumRecords request to a VTab returns the number of records.

The ReturnValue request returns an object matching the class of the field value. You can force the return value to be of the Number or String class by using ReturnValueNumber or ReturnValueString requests against a VTab. Both requests accept two parameters: a field object and a record number.

The Identify window is used to display a specific record. In the following code segment, the user selects a row from a table displayed in the Identify window.

09-05.AVE

```
' Associate this script to the apply event
' of a tool for table documents.
theTable = av.GetActiveDoc
theVTab = theTable.GetVTab
rowNum = theTable.GetUserRow
recNum = theTable.ConvertRowToRecord(rowNum)
theVTab.Identify (recNum,"Identify Results")
```

Editing Records

You can change the value of a field or add new records to the VTab. When the VTab is edited, the physical data source is modified. Before editing a VTab, always verify whether it can be edited by using the following requests.

```
' These requests return a Boolean object.
allowedToEdit = myVTab.CanEdit
canModify = myVTab.IsEditable
canAddToIt = myVTab.CanAddRecord
```

When setting a field value, you will generally want to set it for all records or a certain group of records. You can step through all or a selected group of records, and set a field value for each. The following statements show how to step through records.

```
' Step through selected records.
for each recNum in theVTab.GetSelection
    theVTab.SetValue (aField, recNum, valueObject)
end
'
' Step through all records.
```

```
for each recNum in theVTab
   theVTab.SetValue (aField, recNum, valueObject)
end
```

If you are computing a field value for all or a group of records, it is easier to use the Calculate request. The Calculate request accepts a calculation string and a field to hold the results. The calculation string is the same expression that you would enter in ArcView's calculator. In the following code segment, the proportion of the population over age 18 is computed. The Calculate request applies to the selected records; if no records are selected, the request applies to all records.

```
' Calculate the proportion of the
' population over 18 for all records.
theVTab.Calculate
("100*([POP_TOTAL]-[POP_18])/[POP_TOTAL]",
theVTab.FindField("PCT_18") )
```

To add a record, use the AddRecord request. The request adds a null record. Values can then be set for each field as shown in the following code segment.

```
newRecNum = theVTab.AddRecord
theVTab.SetValueString
(countyField, newRecNum, "FAIRFAX")
```

The SetValue request's third parameter requires an object that matches the field type. In order for the new value to be a Number or String class, use the SetValueNumber or SetValueString request. Both requests accept three parameters: a field object, a record number, and the new value.

Accessing an SQL Database

The access procedure for SQL databases is slightly different from the procedure for file-based data sources. However, once you have created the VTab object for an SQL database, the VTab can be used as discussed earlier in this chapter. This VTab is not editable, however. The following steps describe how to connect to an SQL database and create a VTab object.

1. Verify that the SQL is available by executing the Avenue statement shown below.

```
isThereSQL = SQLCon.HasSQL
```

An SQLCon object is the connection to an SQL database. The HasSQL request to the SQLCon class returns a Boolean oject. The Boolean object is set to True if the SQL feature is available in your ArcView version; otherwise, its value is False.

2. Connect to a database by creating an SQLCon object. You can either get the list of available databases or search for a specific one as shown in the following code segment.

CODE   09-06.AVE

```
' If Oracle server is not available, then
' display a list of available databases to
' the user for selection.
' In Windows, you must run this script from
' a menu item or button.
mySQLConnection = SQLCon.Find ("oracle")
if (mySQLConnection = nil) then
    ' Find failed.
    listSQLCon = SQLCon.GetConnections
    if (listSQLCon.Count = 0) then
      MsgBox.Error
```

```
        ("Unable to find any SQL connection",
        "")
        exit
    else
        mySQLConnection = MsgBox.Choice (listSQLCon,
        "Select a database:", "")
        if (mySQLConnection = nil) then
          exit
        end
    end
end
```

The Find request accepts a string parameter as the database name. If you are not certain about the database name, send a GetConnections request for a list of all available database connections.

3. Log into the server by sending a Login request to your SQLCon object. The Login request needs a parameter as the login string. The code line below includes the login to the Oracle database with the user name "system," and the password "manager."

```
mySQLConnection.Login ("system/manager")
```

4. Create a VTab object based on an SQL query. In the following code segment, a VTab is created for properties valued at over $50,000.

```
anSQLQuery = "select * from property"++
"where land_value > 50000.00"
mySQLVTab = VTab.MakeSQL (mySQLConnection,
anSQLQuery)
```

ArcView stores the definition of the SQL query and automatically reconnects to the database every time the project is opened.

The VTab from the SQL database is in a read-only state. However, you can edit the SQL data by issuing Insert, Delete or Update statements through the ExecuteSQL request. You can also create database objects such as tables with this request. The code segment in the following example updates a table in an SQL database.

```
dmlString = "update process_results"++
"set outcome =  'SUCCESS', comp_tms = sysdate"++
"where id ="++idNum.AsString
executed = mySQLConnection.ExecuteSQL (dmlString)
```

The ExecuteSQL request returns a Boolean object set to True if execution did not cause database errors.

Using ArcStorm

ArcStorm (Arc Storage Manager), ARC/INFO's newest spatial database manager, was introduced with ARC/INFO version 7. Avenue's Librarian class is the equivalent of an ArcStorm database in ARC/INFO. Librarians contain zero or more libraries, libraries hold zero or more layers, and layers hold zero or more data sources. Each data source has a feature class such as line, polygon, or point. The data source can become an ArcView theme.

For example, the state of Virginia could be an ArcStorm database, and each county a library in the database. The layers in each library could be census tracts, census blocks, zip codes, major highways, and streams. The tract and block layers would contain source data for line and polygon feature classes. The zip code layer would have a point feature class as the centroid of each area, in addition to line and polygon features. Finally, the line feature class would be the only data source for the highway and stream layers.

Use the following steps to access a data source from an ArcStorm database.

1. Select the ArcStorm database by creating a Librarian object. In the code segment below, the user is presented with a list of librarians from which to select.

 09-07.AVE

```
' This request returns a list of accessible
' librarians in the $ARCHOME\arcstorm directory.
listLibrarians = Librarian.ReturnLibrarians
if (listLibrarians.Count = 0) then
   MsgBox.Error
   ("Unable to find any ArcStorm databases",
   "")
   exit
end
myLibrarian = MsgBox.Choice (listLibrarians,
"Select an ArcStorm database:", "")
```

If you know the database name, you can create the Librarian object by using the Make request. The code line below creates a Librarian object for the Virginia ArcStorm database.

```
' ARCHOME\arcstorm directory is assumed
' if a path is not provided.
myLibrarian = Librarian.Make ("virginia")
```

2. Select the library by creating a Library object. In the code segment below, the user is presented with a list of libraries from which to select.

09-08.AVE

```
listLibraries = myLibrarian.ReturnLibraries
if (listLibraries.Count = 0) then
    MsgBox.Error ("Unable to find any" ++
    "libraries in" ++ MyLibrarian.GetName ++
    "database.",
    "")
    exit
end
myLibrary = MsgBox.Choice (listLibraries,
"Select a library:", "")
```

If the library name is known, you can create the Library object by using the Make request. In the following code segment, a Library object is created for the Fairfax County library.

```
' Verify whether Fairfax library exists.
foundIt = Library.Exists ("fairfax")
if (Not foundIt) then
    myLibrary = Library.Make ("fairfax", myLibrarian)
end
```

3. To select a layer in the library, create a Layer object. In the code segment below, the user is presented with a list of layers from which to select.

09-09.AVE

```
listLayers = myLibrary.ReturnLayers
if (listLayers.Count = 0) then
    MsgBox.Error ("Unable to find any layers in"++
    myLibrary.GetName++"library",")
    exit
end
myLayer=MsgBox.Choice (listLayers,
"Select a layer:", "")
```

4. A layer can have several data sources, or one data source for each feature class. Access the data source by creating a SrcName object. In the following code segment, the user is requested to select a feature class to add to a view document.

09-10.AVE

```
listSrcNames = Layer.ReturnSrcNames
(myLayer.GetFullName)
if (listSrcNames.Count = 0) then
    MsgBox.Error ("Unable to find any source in"++
    myLayer.GetFullName, "")
    exit
end
listFeature = List.Make
for each source in listSrcNames
    ' Prepare a list of feature classes.
    listFeature.Add (source.GetSubName)
end
aFeature = MsgBox.ChoiceAsString
(listFeature, "Select a feature class:", "")
if (nil = aFeature) then
    exit
end
mySrcName = SrcName.Make
(myLayer.GetFullName++aFeature)
```

The GetFullName request to a Layer object returns a string in the following format: LibrarianName.LibraryName.LayerName. The GetSubName request to a SrcName object

returns the feature class name. The Make request for a Src-
Name object needs one string parameter; the string
parameter depends on the source type, such as ArcStorm,
coverage or shape file. Formats for the string parameter by
source appear below.

```
Source: ArcStorm
Parameter: LibrarianName.LibraryName.LayerName FeatureClass
Source: coverage
Parameter: FullPath\CoverageName FeatureClass
Source: shape file
Parameter: FullPath\ShapeFileName
```

Adding the SrcName object to a view document and
accessing the object's attribute data are easy, as shown
in the following code segment.

```
newTheme = Theme.Make (mySrcName)
aView.AddTheme (newTheme)
newVTab = newTheme.GetFTab
```

The GetFTab request to a feature theme returns an
FTab object as the attribute data.

If you have a large library, you can improve ArcView's
response time by establishing an area of interest
(AOI). Defining an AOI constrains which tiles of the
library display are in the view. Setting the area of interest for a view sets the spatial extent for all library
themes in the view. The AOI can be set by sending the
SetAOI request to a theme or view object. This request
requires a rectangle as its parameter, which must be in
map units and define the extent of AOI.

Programming Chart Documents

Charts provide a powerful visual presentation of attribute data. This chapter illustrates the creation, access and manipulation of chart documents through an application exercise that creates two charts of U.S. counties. One of the charts compares the number of married and divorced individuals, and the other shows median rental cost.

Scripts for Creating Charts

Appearing below is the entire Avenue script to create the two charts. Once developed, the script can be executed from the project window or through an interface control item such as a button.

Before executing this script, create a view named *USA* and add the *counties.shp* shape file as a theme. This shape file is in the *usa* directory of the ArcView data. Change the theme's name to *county* and join its table to *codemog.dbf* using the Fips field. The *codemog.dbf* file is in the *usa\table* directory of ArcView data. Next, select a few counties so that charts are not generated for all counties.

 10-01.AVE

```
' Retrieve the basic information.
thisProject = av.GetProject
theView = thisProject.FindDoc ("USA")
if (nil = theView) then
   MsgBox.Error
   ("Unable to find the view document for USA",
   "")
   exit
end
theTheme = theView.FindTheme ("County")
theVTab = theTheme.GetFTab
'
' Retrieve the required fields for charts.
fieldList1 = { theVTab.FindField("MARRIED"),
theVTab.FindField("DIVORCED") }
fieldList2 = { theVTab.FindField("MEDIANRENT") }
nameField = theVTab.FindField("NAME")
if(fieldList1.Get(0) = nil) then
```

```
    MsgBox.Error("Unable to find fields",)"
    exit
end
'
' Create the charts and set their properties.
pieChart = Chart.Make (theVTab, fieldList1)
pieChartWin = pieChart.GetWin
pieDisplay = pieChart.GetChartDisplay
barChart = Chart.Make (theVTab, fieldList2)
barChartWin = barChart.GetWin
barDisplay = barChart.GetChartDisplay
'
pieDisplay.SetType (#CHARTDISPLAY_PIE)
pieDisplay.SetStyle (#CHARTDISPLAY_VIEW_CUMULATIVE)
pieChart.SetName ("Marital Status")
pieChart.SetSeriesFromRecords (True)
pieChart.SetRecordLabelField (nameField)
pieChart.GetTitle.SetVisible (False)
if (pieDisplay.IsOK.Not) then
    proceed = MsgBox.YesNo
    ("Pie chart may have an inconsistency."
    +NL+"Status:"++pieDisplay.GetStatus
    +NL+"Do you want to continue?", "", False)
    if (Not proceed) then
      exit
    end
end
'
barDisplay.SetType (#CHARTDISPLAY_BAR)
barDisplay.SetStyle (#CHARTDISPLAY_VIEW_SIDEBYSIDE)
barChart.SetName ("Rent")
barChart.SetSeriesFromRecords (True)
barChart.GetChartLegend.SetVisible (False)
barChart.GetYAxis.SetAxisVisible (False)
barChart.GetYAxis.SetLabelVisible (False)
barChart.GetYAxis.SetTickLabelsVisible (False)
barChart.GetXAxis.SetName ("Dollars")
barChart.GetXAxis.SetLabelVisible (True)
barChart.GetXAxis.SetMajorGridSpacing (100)
```

```
barChart.GetXAxis.SetBottom (True)
barChart.GetTitle.SetName ("Median Monthly Rent")
if (barDisplay.IsOK.Not) then
    proceed = MsgBox.YesNo
    ("Bar chart may have an inconsistency"
    +NL+"Status:"++barDisplay.GetStatus
    +NL+"Do you want to continue?", "", False)
    if (Not proceed) then
      exit
    end
end
'

' Add the charts to the project and open them.
thisProject.AddDoc (pieChart)
thisProject.AddDoc (barChart)
' Close all documents first.
thisProject.CloseAll
theView.GetWin.Open
pieChartWin.Open
barChartWin.Open
av.TileWindows
```

Results of executing the above script are illustrated in the following figure.

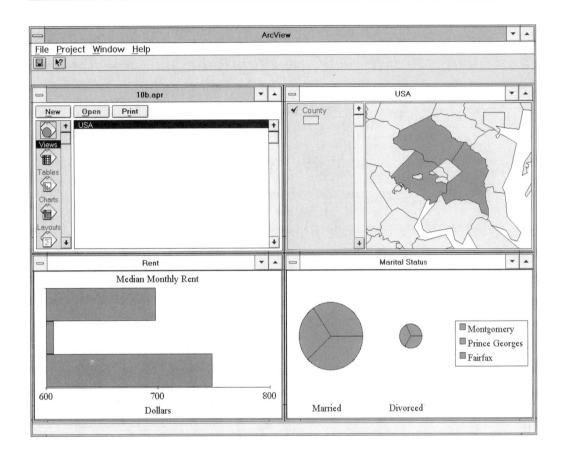

Chart Document

A chart document displays data in a table document as business graphics. Charts use the selected records and fields of a VTab to create graphics. VTabs are discussed further in Chapter 8.

The ChartDisplay, ChartLegend, Title, YAxis, and XAxis classes make up the chart class, which can encompass any, or all, of the five. The five classes are manipulated to create the desired chart. Because charts are documents, you can open, close, and resize the window that holds a chart.

Creating a Chart

Because charts display data in tables, you must already have a table with attribute fields to create a chart. A chart is created by sending the Make request to the chart class. In the following example, the required parameters for the Make request are obtained, and the chart is created.

```
thisProject = av.GetProject
theView = thisProject.FindDoc ("USA")
theTheme = theView.FindTheme("County")
theVTab = theTheme.GetFTab
fieldList1 = { theVTab.FindField("MARRIED"),
theVTab.FindField("DIVORCED") }
'

' Create the chart.
pieChart = Chart.Make (theVTab, fieldList1)
```

The above approach is recommended for static applications in which the table or fields do not change. On the other hand, you can associate the VTab and open a chart property dialog box for the user to select the fields. In the following example, a chart is created after the user clicks on the OK button of the Chart Property dialog box.

```
thisProject = av.GetProject
theView = thisProject.FindDoc ("USA")
theTheme = theView.FindTheme("County")
theVTab = theTheme.GetFTab
'

' Create the chart.
pieChart = Chart.MakeUsingDialog (theVTab,"Chart")
```

The above approaches can be combined to display a Chart Property dialog box showing selected fields. Create the chart and then open the property dialog box as shown in the code segment below. The figure following the code segment shows the property dialog box.

```
' Create a chart.
pieChart = Chart.Make (theVTab, fieldList1)
' Open property dialog box.
pieChart.Edit
```

Chart property dialog box.

Understanding the ChartDisplay and ChartPart classes is important for programming chart documents. Because ChartDisplay holds the visual representation of the chart, manipulating a ChartDisplay object affects the representation. The following code line shows how to access the display object.

```
aChartDisplay = aChart.GetChartDisplay
```

ChartPart is an item in the chart, such as the title or legend. ChartPart subclasses, including Axis, Title, and

ChartLegend, are specific items on the chart and can be directly manipulated. The code segment below is an example of how to manipulate part of a chart.

```
barChart.GetTitle.SetName
("Median Monthly Rent")
```

Setting Chart Type and Style

Chart formats are defined by type and style. A suitable format for the underlying data can enhance presentation effectiveness. ArcView provides six chart types: area, column, bar, line, pie, and XY scatter charts. All types are offered in different styles. The following requests to a ChartDisplay object set the type and style of a chart.

```
aChartDisplay.SetType ( aChartType )
aChartDisplay.SetStyle ( aChartStyle )
```

SetType and SetStyle parameter enumerations are listed below.

Type enumerations (aChartType)

❏ #CHARTDISPLAY_AREA

❏ #CHARTDISPLAY_COLUMN

❏ #CHARTDISPLAY_BAR

❏ #CHARTDISPLAY_LINE

❏ #CHARTDISPLAY_PIE

❏ #CHARTDISPLAY_XYSCATTER

Style enumerations (aChartStyle)

- ☐ #CHARTDISPLAY_VIEW_CUMULATIVE

- ☐ #CHARTDISPLAY_VIEW_RELATIVE

- ☐ #CHARTDISPLAY_VIEW_SIDEBYSIDE

The following code segment is an example of setting chart type and style.

```
barChart = Chart.Make (theVTab, fieldList2)
barDisplay = barChart.GetChartDisplay
barDisplay.SetType (#CHARTDISPLAY_BAR)
barDisplay.SetStyle (#CHARTDISPLAY_VIEW_SIDEBYSIDE)
```

Through ChartDisplay, other visual properties such as markers or symbols can be set in line and scatter charts. Exercise care when setting ChartDisplay properties because not all combinations are valid. When properties are inconsistent, charts are not displayed. One way to prevent inconsistencies is to create the desired chart through ArcView. Then use Get requests to determine the properties set by ArcView. In the example below, the type and style of an existing chart are examined. The two figures following the script show the expected results.

```
myChart = av.GetProject.FindDoc ("Test Chart")
myChartDisplay = myChart.GetChartDisplay
MsgBox.Info ("Type:"
++myChartDisplay.GetType.AsString,
"Test Chart")
"MsgBox.Info ("Style:"
"++myChartDisplay.GetStyle.AsString,
"Test Chart")
```

Chart type display.

Chart style display.

Avenue provides a request to check for inconsistencies in a chart. The code segment below shows how to use the IsOK request to determine inconsistencies. If inconsistencies are found, the GetStatus request is used to identify them.

```
if (pieDisplay.IsOK.Not) then
   proceed = MsgBox.YesNo
   ("Pie chart may have an inconsistency"
   +NL+"Status:"++pieDisplay.GetStatus
   +NL+"Do you want to continue?", "", False)
   if (Not proceed) then
     exit
   end
end
```

Setting Chart Properties

Chart properties are set by manipulating legend, title, and axis objects. In addition, other properties such as name and data labels are accessed through the chart object. The following requests deal with the properties of a chart object.

```
groupByField = aChart.IsSeriesFromRecords
aChart.SetSeriesFromRecords (True)
```

A chart may be grouped by fields or records. In turn, series are set to records or fields. A group is a set of chart graphical elements that are brought together. A series is a set of chart graphical elements that are drawn once in each group. For example, in the "Marital Status" chart, "married" and "divorced" are groups, while each county forms a series. The following code lines access and set the label of a group identified by *aGroupNumber.*

```
aGroupLabel = aChart.GetGroupLabel (aGroupNumber)
aChart.SetGroupLabel (aGroupNumber, aGroupLabel)
```

The aGroupNumber object refers to one group by its sequence number; zero refers to the first group, and aGroupLabel is a string object.

```
aSeriesLabel = aChart.GetSeriesLabel (aSeriesNumber)
aChart.SetSeriesLabel (aSeriesNumber, aSeriesLabel)
```

The aSeriesNumber object refers to a series by its sequence number; zero refers to the first series and aSeriesLabel is a string object. If groups or series are from fields, the request to set a label also sets an alias for the field.

To access chart parts such as title or legend, the appropriate object must be retrieved. The following requests retrieve chart parts.

```
aChartLegend = aChart.GetChartLegend
aChartTitle = aChart.GetTitle
aChartXAxis = aChart.GetXAxis
aChartYAxis = aChart.GetYAxis
```

A chart part can be manipulated once you have its object. XAxis and YAxis are subclasses of Axis; Axis provides most of the requests for manipulating either axis. The code segment below is an example of setting properties for chart parts.

```
' Do not show the legend and Yaxis.
barChart.GetChartLegend.SetVisible (False)
barChart.GetYAxis.SetAxisVisible (False)
barChart.GetYAxis.SetLabelVisible (False)
' Set Xaxis label to Dollars.
barChart.GetXAxis.SetName ("Dollars")
barChart.GetXAxis.SetLabelVisible (True)
' Set the chart title.
barChart.GetTitle.SetName
("Median Monthly Rent")
```

Various requests for each chart part are listed below.

```
' Set the visibility of an axis.
anAxis.SetAxisVisible (True)
' Set maximum and minimum values.
anAxis.SetBoundsMax (maxNumber)
anAxis.SetBoundsMin (minNumber)
anAxis.SetBoundsUsed (True)
' Set and display grid lines.
anAxis.SetMajorGridSpacing (aNumber)
anAxis.SetMajorGridVisible (True)
'
' Get or set the location of a chart legend.
aChartLoc = aChartLegend.GetLocation
aChartLegend.SetLocation (aChartLoc)
' Do not display a chart legend.
aChartLegend.SetVisible (False)
'
' Set the title.
aChartTitle.SetName (aString)
' Get or set the location of a chart title.
aChartLoc = aChartTitle.GetLocation
```

```
aChartTitle.SetLocation (aChartLoc)
' Display a chart title.
aChartTitle.SetVisible (True)
```

The aChartLoc enumerations are as follows.

❐ #chartDISPLAY_LOC_BOTTOM

❐ #chartDISPLAY_LOC_LEFT

❐ #chartDISPLAY_LOC_RELATIVE

❐ #chartDISPLAY_LOC_RIGHT

❐ #chartDISPLAY_LOC_TOP

Chart is a subclass of Doc. A chart object inherits all the valid requests for Doc. Some of the most commonly used requests appear in the code segment below.

```
' Document window is needed to open or
' close a document.
aChartWin = aChart.GetWin
aChartWin.Open
aChartWin.Close
'
' Document name appears in the project window and
' document title bar.
aChart.SetName ("my Chart")
```

Working with Data Elements

In most applications, tabular data is used to create a chart. However, in some cases you may wish to access data from a chart. Once a chart is created in an application, the user can click on a customized button to calculate a series of complex statistics. For example, in a redistricting application a bar chart might show the population's racial composition for an arbitrary set of districts. The user could then click on a button to compute the standard deviation for all districts in the chart.

The following data elements can be accessed: records in the chart, fields used in the chart, a record identified by the user, and a record with a matching string of characters. The data elements are described in the following sections.

Accessing Records

The chart document displays selected records in a VTab. If no records are selected, all records are displayed in a chart. To access the records from a chart, the user must access its VTab and then the selected rows in the VTab. The code segment below demonstrates this approach.

```
theVTab = myChart.GetVTab
selectedRecords = theVTab.GetSelection
if (selectedRecords.Count = 0) then
' No records were selected.
' Therefore, all records are shown in the chart.
Else
   for each recNumber in selectedRecords
    ' The record referenced by the recNumber variable
    ' is part of the chart.
   end
end
```

The GetSelection request returns a bit map object. A bit map is an ordered, fixed-size list of Boolean objects. Each Boolean object represents a record with a matching index number in the VTab. If the Boolean object is set to True, the record is selected. A list of useful requests for a bit map object appears below.

```
' Clear all bits by assigning a False value.
aBitmap.ClearAll
'

' Toggle True and False bits.
aBitmap.Not
'

' Set all bits to True.
aBitmap.SetAll
'

' Return number of bits that are set.
countSet = aBitmap.Count
'

' Return the size of a bit map.
size = aBitMap.GetSize
'

' Evaluate a bit map at a bit location.
isSet = aBitmap.Get (aBitLocation)
```

Accessing Fields

A list of fields used in the chart is returned by the Get-Fields request to the chart object. The code segment below shows how to check if the *POP18* field is used in a chart.

```
fieldsList = myChart.GetFields
for each aField in fieldsList
   If (aField.GetName = "POP18") then
   ' The field has been found.
   end
end
```

Identifying Records

ArcView provides an Identify tool button for chart documents. With the use of this tool, ArcView displays the fields of a particular record that pertain to a chart data element. The Identify tool button can also be used to perform other tasks with the identified record. In the code segment below, the GetRecord-FromClick request accepts a mouse click from the user. If the user clicks on a data element, the request returns the associated record number.

```
theVTab = myChart.GetVtab
recNumber = myChart.GetUserRecord
if (recNumber >= 0) then
   population = theVTab.ReturnValue
   (popField,recNumber)
end
```

Finding Records by Matching Strings

The Find request can search for a matching string in any of the character fields of the associated VTab. If a record is found, the request returns a record number. A nil object is returned if a record is not found. The format for this request is shown below.

```
recNumber = myChart.Find (aString)
```

Printing Charts

Printing a chart is simple. The only task involved is sending the Print request to a chart object as shown in the code line below.

```
myChart.Print
```

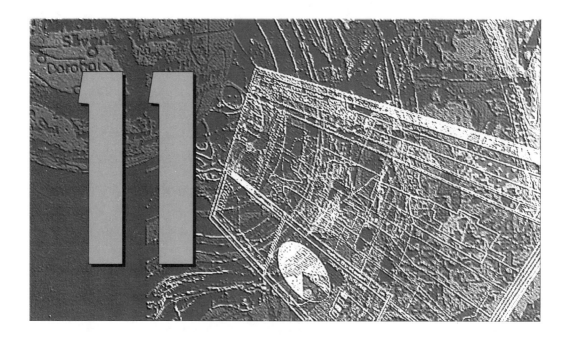

Programming Layout Documents

Users prepare hardcopy maps in ArcView by using layout documents. A layout document is a map that can contain views, tables, charts, imported graphics, and graphic primitives. It can also contain cartographic elements such as scale bars or north arrows.

Avenue can automate the preparation of a layout document. You can write an Avenue script to guide a user through generating a custom map or producing a standard layout. For example, map standards in your organization may require that the north arrow appear above the scale bar at the bottom center of the printed page. Avenue statements provide the flexibility to pick from different north arrow or scale bar styles, while consistently placing them in the center at the bottom of the sheet. In another situation your organization could use a specific style of north arrow regardless of its location on the sheet. An Avenue script can provide flexibility of location while retaining a predefined north arrow style.

This chapter discusses how to use Avenue to create and modify a layout document. A script will be created and then associated to a customized button of a view document interface. Clicking the button generates a layout as shown in the following figure. All the elements in this layout are created through Avenue. You can save some programming effort when you create a layout interactively, store it on disk as a template, and reuse it when needed. This approach is explained later in this chapter.

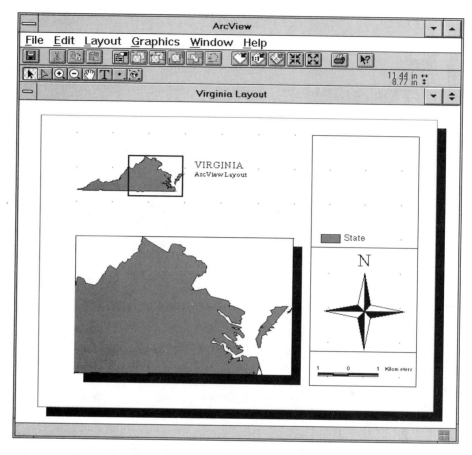

Layout document generated by an Avenue script.

Layout Document Script

The entire script follows. Each segment of the script is explained afterward.

11-01.AVE

```
' CreateAMap
' This script is attached to a customized
' button in the view document interface.
' Executing this script will create a standard
' layout document from the current view.
'
' First create a new layout.
'
thisProject = av.GetProject
thisView = av.GetActiveDoc
stdLayout = Layout.Make
stdLayout.SetName (thisView.GetName++"Layout")
stdLayoutDpy = stdLayout.GetDisplay
stdLayoutGL = stdLayout.GetGraphics
'
' Set the layout page properties to
' landscape 11 x 8.5 with an all-around
' 0.00 inch margin.
'
marginRect = Rect.MakeXY ( 0.0, 0.0, 0.0, 0.0)
stdLayoutDpy.SetUnits (#UNITS_LINEAR_INCHES)
stdLayoutDpy.SetMargin (marginRect)
stdLayoutDpy.SetMarginVisible (True)
stdLayoutDpy.SetPageSize (Point.Make(11.0,8.5))
stdLayoutDpy.SetGridActive (False)
stdLayoutDpy.SetGridVisible (True)
stdLayoutDpy.SetGridMesh (Point.Make(1.0,1.0))
'
' Requests that place graphic objects on the
' layout require X and Y coordinates. The
' coordinates are based on the Display frame
' and not the PageDisplay.
' In order to use the coordinates of
' PageDisplay, the
' lower left X and Y for the PageDisplay is
' obtained. In the next code line this
' coordinate is stored in
' the oPt (origin point) object.
'
```

```
oPt = stdLayoutDpy.ReturnMarginExtent.ReturnOrigin
'
' Add a title and a subtitle to the page.
' Use Times Roman font and sizes of 24 and 18.
'
titleSymbol = TextSymbol.Make
titleSymbol.SetFont (
Font.Make("Times New Roman","Normal") )
titleSymbol.SetSize (24) 'size is in points
subtitleSymbol = titleSymbol.Clone
subtitleSymbol.SetSize (18)
allTitles = MsgBox.MultiInput (
"Please enter","",
{"Title:","Subtitle:"},
{"",""})
if (allTitles <> nil) then
   pTitle = GraphicText.Make (
   allTitles.Get(0), oPt+Point.Make(4.25,7.0) )
   pTitle.SetSymbols ({titleSymbol})
   pTitle.SetAngle (0)
   stdLayoutGL.Add (pTitle)
   subtitle = GraphicText.Make (
   allTitles.Get(1), oPt+Point.Make(4.25,6.75) )
   subtitle.SetSymbols ({subtitleSymbol})
   subtitle.SetAngle (0)
   stdLayoutGL.Add (subtitle)
else
   exit
end
'
' Add view frames. The primary view frame
' shows the active view at its current extent,
' while the locator view frame shows the
' active frame zoomed to its extent with
' a rectangle depicting the area of primary view.
'
' The primary view is placed at point 1,1 with
' a shadow box.
'
```

```
vFill = RasterFill.Make
vFill.SetStyle (#RASTERFILL_STYLE_SOLID)
vFill.SetColor (Color.GetWhite)
vFill.SetOutlined (true)
vFill.SetOLColor (Color.GetBlack)
vRect = Rect.Make (
oPt+Point.Make (1.0,1.0),
Point.Make (6.0,4.0) )
vRectGr = GraphicShape.Make (vRect)
vRectGr.SetSymbol (vFill)
stdLayoutGL.Add (vRectGr)
vFrame = ViewFrame.make (vRect)
vFrame.SetSymbol (vFill)
vFrame.SetView (thisView,true)
vFrame.SetScalePreserved (false)
stdLayoutGL.Add (vFrame)
'

' Add a shadow box to the primary view.
'

sFill = vFill.Clone
sFill.SetColor (Color.GetBlack)
sRect = vFrame.GetBounds
shadowGr = GraphicShape.Make (sRect)
shadowGr.Offset (Point.Make(0.25,-0.25))
shadowGr.SetSymbol (sFill)
stdLayoutGL.UnselectAll
shadowGr.SetSelected (true)
stdLayoutGL.Add (shadowGr)
stdLayoutGL.MoveSelectedToBack
stdLayoutGL.UnselectAll
'

' Add the locator map at 1,5.5 inches.
' The locator map is based on a cloned view that
' displays the view's full extent.
'

lRect = Rect.Make (
oPt+Point.Make(1.0,5.5),
Point.Make (3.0,2.5) )
lFrame = ViewFrame.Make (lRect)
```

```
lFrame.SetSymbol (vFill)
fullView = thisView.Clone
fullView.GetDisplay.SetExtent
(fullView.ReturnExtent)
fullView.SetName("Full View")
lFrame.SetView (fullView, True)
lFrame.SetScalePreserved (False)
stdLayoutGL.Add (lFrame)
'

' Draw a box in the locator map to show
' the extent of the primary view. This box is
' actually drawn on the view display and
' seen on the layout document.
'

boxFill = vFill.Clone
boxFill.SetStyle (#RASTERFILL_STYLE_EMPTY)
boxFill.SetOutlined (True)
boxFill.SetOlColor (Color.GetBlack)
boxFill.SetOlWidth (4)
vFrameExtent = thisView.GetDisplay.ReturnVisExtent
lBox = GraphicShape.Make (vFrameExtent)
lBox.SetSymbol (boxFill)
fullViewGL = fullView.GetGraphics
fullViewGL.Add (lBox)
'

' Draw boxes to hold the scale bar,
' north arrow, and legend. AddBatch is
' used to add the remaining items to
' the layout.
'

aPen = BasicPen.Make
aPen.SetColor (Color.GetBlack)
infoBox = Polygon.Make ( { {
oPt+Point.Make (7.50,0.75),
oPt+Point.Make (7.50,8.00),
oPt+Point.Make (10.5,8.00),
oPt+Point.Make (10.5,0.75) } } )
infoBoxGr = GraphicShape.make (infoBox)
infoBoxGr.SetSymbol (aPen)
```

```
stdLayoutGL.AddBatch (infoBoxGr)
'

line1 = Line.Make (
oPt+Point.Make (7.50,1.75),
oPt+Point.Make (10.5,1.75) )
line1Gr = GraphicShape.Make (line1)
line1Gr.SetSymbol (aPen)
stdLayoutGL.AddBatch (line1Gr)
'

line2 = Line.Make (
oPt+Point.Make (7.50,4.75),
oPt+Point.Make (10.5,4.75) )
line2Gr = GraphicShape.Make (line2)
line2Gr.SetSymbol (aPen)
stdLayoutGL.AddBatch (line2Gr)
'

' Add scale bar.
'

sbRect = Rect.Make (
oPt+Point.Make (7.7,1.0),
Point.Make (2.45,0.24) )
sbFrame = ScalebarFrame.Make (sbRect)
sbFrame.SetUnits (#UNITS_LINEAR_KILOMETERS)
sbFrame.SetStyle (#SCALEBARFRAME_STYLE_ALTFILLED)
sbFrame.SetViewFrame (vFrame)
sbFrame.SetInterval (1)
sbFrame.SetIntervals (1)
stdLayoutGL.AddBatch (sbFrame)
'

' Add north arrow.
'

naRect = Rect.Make (
oPt+Point.Make (7.9, 1.8),
Point.Make (2.75,3.0) )
naGr = NorthArrow.Make (naRect)
' Retrieve a predefined north arrow from
' the north.def file.
northArrowFile = FileName.Make
("d:\progs\esri\arcview\ver30\arcview\etc\north.def")
```

```
northArrowODB = ODB.Open (northArrowFile)
if (nil = northArrowODB) then
   MsgBox.Error
   ("Unable to open north.def object database",
   "")
   exit
end
northArrowList = northArrowODB.Get(0)
anArrow = northArrowList.Get(0)
naGr.SetArrow (anArrow)
stdLayoutGL.AddBatch (naGr)
'
' Add the legend.
'
lgRect = Rect.Make (
oPt+Point.Make (7.75,4.85),
Point.Make (5.65,0.30) )
lgFrame = LegendFrame.Make (lgRect)
lgFrame.SetViewFrame (vFrame)
stdLayoutGL.AddBAtch (lgFrame)
'
' End the AddBatch.
'
stdLayoutGL.EndBatch
'
' Open the layout document.
stdLayoutWin = stdLayout.GetWin
stdLayoutWin.Open
stdLayoutDpy.ZoomToPage
```

The preceding script should be associated with a customized button of the view document interface.

Layout Documents

Layout is a specialized Doc class and can be manipulated in the same way as other documents. This section discusses how to create a new layout, set properties, and display the layout.

Creating a Layout

As seen in the following Avenue statement, creating a layout document requires only a Make request sent to the Layout class. The newly created layout document is automatically added to the project file.

```
stdLayout = Layout.Make
```

Similar to other documents, a layout document has display and window objects. Another important object is the graphic list. The graphic list for a layout document maintains all elements added to the document, such as title text, the north arrow, and frames.

Retrieving the display, window, and graphic list objects should be done immediately after creating the layout document. The code segment below shows how the objects are retrieved.

```
' Retrieve the layout display object.
stdLayoutDpy = stdLayout.GetDisplay
' Retrieve the layout document window.
stdlayoutWin = stdLayout.GetWin
' Retrieve the layout graphic list.
stdLayoutGL = stdLayout.GetGraphics
```

The display object is required to manipulate document contents, such as pan, zoom and page size. The window object is for document management, for example, to open and maximize the document.

Layout Window

Once the layout is created, you can access its window object in order to display the layout and ask user approval before sending it to the printer. In the following code segment, the layout is opened, and the user is asked if the layout should be printed.

```
' Retrieve the window object.
stdLayoutWin = stdLayout.GetWin
' Open layout if closed.
if (stdLayoutWin.IsOpen.Not) then
   stdLayoutWin.Open
end
' Maximize layout.
stdLayoutWin.Maximize
' Ask user whether to print.
goAhead = MsgBox.YesNo ("Print this layout?",
"", true)
if (goAhead) then
   stdLayout.Print
else
   MsgBox.Info ("Print canceled.","")
end
' Close the layout document.
stdLayoutWin.Close
```

Setting Layout Properties

The main properties of a layout document are related to layout components such as page size and margins. Assuming that the printer has been configured in Arc-View, you can simply set the page size and margins to the printer's default values. Use the following requests to obtain the printer default values set in the current configuration.

```
' Both requests are made to the
' layout display object.
myLayoutDpy.SetUsingPrinterPageSize (true)
myLayoutDpy.SetUsingPrinterMargins (true)
```

Page properties can be set directly by sending the following requests to the layout display object.

```
' Set page units to inches.
stdLayoutDpy.SetUnits (#UNITS_LINEAR_INCHES)
' Set to landscape letter size.
' Set page by sending a point equivalent
' to the page size.
stdlayoutDpy.SetPageSize (11@8.5)
' Set margins to 0.25 inches all around.
' Set the margin by sending a rectangle
' equivalent to the margins.
marginRect = Rect.MakeXY (0.25, 0.25, 0.25, 0.25)
stdlayoutDpy.SetMargin (margin.Rect)
' Verify that margin lines are displayed.
stdLayoutDpy.SetMarginVisible (True)
```

In the preceding code segment, the page measurement unit is set by the enumeration value, #UNITS_ LINEAR_INCHES. Additional unit enumerations appear below.

❒ #UNITS_LINEAR_CENTIMETERS

❒ #UNITS_LINEAR_DEGREES

❒ #UNITS_LINEAR_FEET

❒ #UNITS_LINEAR_KILOMETERS

❒ #UNITS_LINEAR_METERS

❐   #UNITS_LINEAR_MILES

❐   #UNITS_LINEAR_MILLIMETERS

❐   #UNITS_LINEAR_NAUTICALMILES

❐   #UNITS_LINEAR_UNKNOWN

❐   #UNITS_LINEAR_YARDS

Grid is another page property. A grid is often useful when a user is interactively creating the layout document because it provides a visual aid for placing layout components. The grid snap also ensures that components are placed in exact locations. However, when preparing a layout through Avenue, a grid system is not required because components can be located at precise locations. The grid can nonetheless provide a visual aid to show that each component is placed in relation to another. In the following code segment, the layout displays a one-inch grid mesh without snapping.

```
' Grid properties are set by sending
' requests to the layout display object.
'
' Set the grid mesh to one-inch intervals;
' set the mesh in x and y directions by sending
' an equivalent point.
stdLayoutDpy.SetGridMesh (1@1)
' Display the grid.
stdLayoutDpy.SetGridVisible (true)
' Turn off the snap.
stdLayoutDpy.SetGridActive (false)
```

Because layout is a type of document, it has properties similar to other documents. The document name, one of the most useful properties, appears on the title bar of a document window. The document name can also be used to search for a document. In the following code segment, a layout document is named by adding the word "Layout" to the primary view name.

```
stdLayout.SetName (thisView.GetName++"Layout")
```

Using the Graphic List

When a component is placed on the layout, it is added to the layout graphic list. The list is used in accessing and manipulating all layout components.

The GraphicList class is a specialization of the List class. The GraphicList class holds the graphics associated with document display. Whenever a document display is redrawn, the graphic list is used to place all the graphics. The graphics in the list are comprised of mathematical representations of shapes, and become objects of the GraphicShape class. GraphicShape provides the algorithm that allows each shape to draw itself.

The simple sequence of adding graphics to a document involves the following steps: retrieve the mathematical representation of the graphic shape, add it to the graphic list, and trigger a redraw of the document. This sequence is illustrated by the following code segment, in which a polygon is added to the layout document.

```
' Get the layout graphic list by
' sending the GetGraphics request
' to the layout document.
```

```
stdLayoutGL = stdLayout.GetGraphics
' Create a shape by sending a
' Make request to a shape class.
infoBox = Polygon.Make ( { {
oPt+Point.Make (7.50,0.75),
oPt+Point.Make (7.50,8.00),
oPt+Point.Make (10.5,8.00),
oPt+Point.Make (10.5,0.75) } } )
' Generate the mathematical
' representation of the shape
' by sending a Make request to
' the GraphicShape class.
infoBoxGr = GraphicShape.make (infoBox)
' Add the graphic shape to the
' graphic list and cause a redraw.
stdLayoutGL.Add (infoBoxGr)
stdLayout.Invalidate
```

When several graphic shapes are added to a graphic list, it is frequently more efficient to use the AddBatch request in conjunction with the EndBatch request. These requests are shown in the statements below.

```
' Simultaneously add several graphic
' shapes to the graphic list.
stdLayoutGL.AddBatch (line1Gr)
stdLayoutGL.AddBatch (line2Gr)
stdLayoutGL.AddBatch (line3Gr)
stdLayoutGL.EndBatch
```

The following code segment erases all layout components.

```
stdLayoutGL = stdLayout.GetGraphics
' Select all graphics.
stdLayoutGL.SelectAll
' Erase all selected graphics.
stdLayoutGL.ClearSelected
```

An object from the GraphicText class is used for adding text to the layout document. In the following code segment, this process is illustrated by adding a title to the layout document.

```
' Get the layout graphic list
' by sending the GetGraphics request
' to the layout document.
stdLayoutGL = stdLayout.GetGraphics
' Get the title from the user.
allTitles = MsgBox.MultiInput (
"Please enter",""
{"Title:","Subtitle:"},
{"",""})
if (allTitles<>nil) then
   ' Generate the mathematical
   ' representation of the text
   ' by sending a Make request to
   ' the GraphicText class.
   pTitle = GraphicText.Make (
   allTitles.Get(0), locationPoint)
   ' Add the graphic text to the list.
   stdLayoutGL.Add (pTitle)
end
```

The GraphicShape and GraphicText classes create the graphics for shapes and text to add to the layout graphic list. Graphical frames are required to add views, charts, or tables. Frames, which can be described as containers for views or other layout components, are discussed in the following sections.

Framing a View Document

Views are placed on a layout document through objects of the ViewFrame class. A view frame is created by sending the Make request to the ViewFrame class. As a parameter, this request requires a rectangle that represents the frame boundary. An existing view document can also be associated to the frame. If a view document is not associated, then the frame is displayed as an empty view frame. In the following code segment, a view frame is created, its properties are set, and the frame is added to the graphic list.

```
' Add a view frame to the stdLayout document.
'
' Set the frame boundary in vRect.
vRect = Rect.Make (
oPt+Point.Make (1.0,1.0),
Point.Make (6.0,4.0) )
' Create the view frame.
vFrame = ViewFrame.make (vRect)
' Assign a view to the frame. The second
' parameter sets the live link; if set to False,
' view frame becomes a snapshot of the view document.
vFrame.SetView (thisView,true)
' By preserving the scale, frame displays the
' view at its current scale;
   ' otherwise the scale is changed to match the extent.
vFrame.SetScalePreserved (false)
' Add the view frame to the graphic list.
stdLayoutGL.Add (vFrame)
' Redraw the layout.
stdLayout.Invalidate
```

The layout document may change the size of the view frame based on the width and height of the associated view. The GetBounds request can be used to obtain the new size. This request is useful when placing another object on the layout based on the view frame. For example, to place a shadow box under the view frame, you need the exact frame size. In the following code segment, a shadow box for the view frame is added to the layout.

```
' Add a shadow box to the vFrame object.
'
' First, create a solid black fill symbol.
sFill = RasterFill.Make
sFill.SetStyle (#RASTERFILL_STYLE_SOLID)
sFill.SetColor (Color.GetBlack)
' Set the shadow box size to the same size as the
' vFrame. GetBounds returns a rectangle.
sRect = vFrame.GetBounds
' Create a mathematical representation of a
' rectangle.
shadowGr = GraphicShape.Make (sRect)
' Move the shadow box by half-inch increments to the right
' and down.
shadowGr.Offset (Point.Make(0.5,-0.5))
' Set the shadow box to a black solid fill symbol.
shadowGr.SetSymbol (sFill)
' Verify that only the shadow box
' is selected in the graphic list.
stdLayoutGL.UnselectAll
shadowGr.SetSelected (true)
' Add shadow box to the layout, and then
' move it to the back of the view frame.
stdLayoutGL.Add (shadowGr)
stdLayoutGL.MoveSelectedToBack
stdLayoutGL.UnselectAll
```

Once a ViewFrame object is placed on the layout, the associated scale bar and legend can be added.

Adding a Scale Bar

A scale bar can be drawn by associating a ScaleBar-Frame object to a view frame and adding it to the graphic list. Send the Make request to the Scalebar-Frame class. This creates a scale bar frame object. The Make request requires a rectangle that sets the scale bar's boundary. The following code segment is an example of how to add a scale bar to a view.

```
' Add a scale bar.
'
' Create a rectangle defining the scale
' bar boundary.
sbRect = Rect.Make (
oPt+Point.Make (7.7,1.0),
Point.Make (2.45,0.24) )
' Create the scale bar object.
sbFrame = ScalebarFrame.Make (sbRect)
' Set scale bar properties.
sbFrame.SetUnits (#UNITS_LINEAR_KILOMETERS)
sbFrame.SetStyle (#SCALEBARFRAME_STYLE_ALTFILLED)
' Associate the bar with the vFrame.
sbFrame.SetViewFrame (vFrame)
' Add the bar to the graphic list.
stdLayoutGL.Add (sbFrame)
```

Several scale bar styles are available through the bar style enumeration. The styles are listed below.

❐ #SCALEBARFRAME_STYLE_ALTFILLED

❐ #SCALEBARFRAME_STYLE_FILLED

❐ #SCALEBARFRAME_STYLE_HOLLOW

❐ #SCALEBARFRAME_STYLE_LINED

❐ #SCALEBARFRAME_STYLE_TEXT

Additional properties of a scale bar can be set through Avenue. These properties are shown in the following code lines.

```
' Set the number of divisions in the left
' interval of the scale bar.
myScaleBar.SetDivisions (aNumber)
'
' Set the interval of the scale bar.
myScaleBar.SetInterval (aNumber)
```

Adding Legends

The view document legend can be displayed on the layout inside a legend frame. An object of the LegendFrame class is created by sending a Make request. The Make request requires a rectangle corresponding to the boundary of the legend frame. A legend frame must be associated to a view frame. The following code segment shows how to add a legend to the layout.

```
' Add a legend.
'
' Create a rectangle for the frame boundary.
lgRect = Rect.Make (
oPt+Point.Make (7.75,4.85),
Point.Make (5.65,0.30) )
' Create a legend frame object.
lgFrame = LegendFrame.Make (lgRect)
' Associate the legend with the view frame.
lgFrame.SetViewFrame (vFrame)
' Add the legend frame to the graphic list.
stdLayoutGL.Add (lgFrame)
```

Adding a North Arrow

The NorthArrow class is a specialized class of GraphicGroup. NorthArrow objects are limited to graphic shapes and graphic text. You can create your own north arrow by creating shape and text graphics and adding them to the NorthArrow graphic group. However, ArcView comes with an ODB (object database) file that defines several types of north arrows. ODBs are discussed in Chapter 12. The following code segment shows how to access the *north.def* file in the *etc* directory to retrieve a NorthArrow object.

```
' Several north arrow shapes are defined
' in the north.def object database file.
northArrowFile = File.Make
("\esri\arcview\etc\north.def")
northArrowODB = ODB.Open (northArrowFile)
' The first object in the ODB is a list of
' stored objects.
northArrowList = northArrowODB.Get(0)
anArrow = northArrowList.Get(0)
naGr = NorthArrow.Make (aRect)
naGr.SetArrow (anArrow)
stdLayoutGL.Add (naGr)
```

Framing a Table or Chart Document

A table or chart document is placed on a layout by adding a DocFrame object to the layout graphic list. The following code segment shows how to create a DocFrame object and add it to the graphic list.

```
' Document frames are created within
' rectangle boundaries.
tableRect = Rect.MakeXY (x1, y1, x2, y2)
tableFrame = DocFrame.Make (tableRect, Table)
```

```
' Associate frame with the table document.
tableFrame.SetFramedDoc (myTable)
' Add frame to the layout graphic list.
myLayoutGL.Add (tableFrame)
'
' A chart document frame is created in
' a similar way.
chartRect = Rect.MakeXY (x1, y1, x2, y2)
chartFrame = DocFrame.Make (chartRect, Chart)
' Associate frame with the chart document.
chartFrame.SetFramedDoc (myChart)
' Add frame to the layout graphic list.
myLayoutGL.Add (chartFrame)
```

The layout does not display the contents of a table or chart document unless the table or chart document is open. A DocFrame object is a type of frame that holds a table or chart on the layout. The following properties can be set for a DocFrame object.

```
' Set presentation quality.
myDocFrame.SetQuality (#FRAME_QUALITY_DRAFT)
' The other enumeration for presentation
' quality is #FRAME_QUALITY_PRESENTATION.
'
' Set the frame refresh property.
myDocFrame.SetRefresh (#FRAME_REFRESH_WHENACTIVE)
' The other enumeration for the refresh
' property is #FRAME_REFRESH_ALWAYS.
```

Framing a Picture

Contents of a graphic file can be placed on the layout by adding a PictureFrame object to the graphic list. A graphic disk file such as an organization's logo or a property photo can be imported into a picture frame. In the following code segment, a rectangle area for a company logo is created.

```
logoRect = Rect.MakeXY (x1, y1, x2, y2)
' Create a picture frame within the logoRect
' boundaries.
logoFrame = PictureFrame.Make (logoRect)
' Import the graphic file into the frame.
logoFrame.SetFileName ("LOGO.TIF".AsFileName)
' Add file to the graphic list.
myLayoutGL.Add (logoFrame)
```

The presentation quality and refresh properties for a picture frame can also be set. The appropriate requests are shown in the following code segment.

```
logoFrame.SetQuality (qualityEnum)
' The qualityEnum could have one of the
' following values: #FRAME_QUALITY_DRAFT
' or #FRAME_QUALITY_PRESENTATION.
'
logoFrame.SetRefresh (refreshEnum)
' The refreshEnum could have one of the
' following values:
' #FRAME_REFRESH_WHENACTIVE
' or #FRAME_REFRESH_ALWAYS.
```

The user can also select the graphic file by browsing through the directories. In the following code segment, a picture frame is created but the user selects the graphic file. The picture frame is selected prior to placing it on the layout so that the user can immediately manipulate its size and location.

```
photoRect = Rect.MakeXY (x1, y1, x2, y2)
photoFrame = PictureFrame.Make (photoRect)
if (photoFrame.Edit (myLayoutGL)) then
    ' Verify that nothing is selected.
    myLayoutGL.UnselectAll
    ' Select the picture frame.
    photoFrame.SetSelected (true)
```

```
' Add the frame to the layout graphic list.
myLayoutGL.Add (photoFrame)
end
```

The Edit request is used to set a frame's properties, including the graphic file for import. The property dialog box, as shown in the following figure, contains an OK and a Cancel button. The Edit request returns True if OK is clicked. If the user does not click on OK, the request returns False.

Picture Frame Properties dialog box.

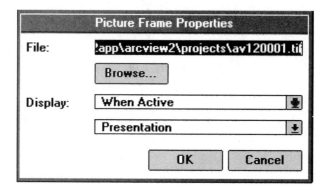

ArcView supports the following graphic formats.

❏   Band Interleave (bil, bip, bsq)

❏   Compuserve (gif)

❏   Encapsulated PostScript (eps)

❏   ERDASgis (lan)

❏   IMPELL (rlc)

❏   Mac Paint (mcp)

❐ Nexpert Object Image (nbi)

❐ PostScript (ps)

❐ Sun Raster (rs)

❐ TIFF (tif)

❐ Windows Bitmap (bmp)

❐ Windows Metafile (wmf)

❐ X-Bitmap (xbm)

Using Templates

Standard layouts can be created, stored on disk as templates, and then used to create new layouts. As shown earlier in this chapter, you can create these layouts with Avenue. Store them as templates using the following code segment.

```
myTemplate = myLayout.AsTemplate
myTemplate.SetName ("My New Template")
TemplateMgr.Add (myTemplate)
```

It may be easier to create the layout and template interactively. New templates are stored in *template.def* file. The following script shows how to access *template.def*, ask the user to select a template by name, and then apply it to a layout.

11-02.AVE

```
' Load Template.
tList = TemplateMgr.GetTemplates
theTemplate = MsgBox.Choice
(tList,"Select a Template","")
theLayout = Layout.Make
theLayout.SetName ("MAP Layout")
' Set page size.
layoutDpy = theLayout.GetDisplay
layoutDpy.SetUnits (#UNITS_LINEAR_INCHES)
layoutDpy.SetUsingPrinterPageSize (false)
layoutDpy.SetPageSize (36@24)
' Apply the template.
theLayout.UseTemplate (theTemplate)
' Fill the template frames.
orderList= {av.GetProject.FindDoc("Virginia")}
theLayout.GetGraphics.FillFrames (orderList)
```

The FillFrames request accepts a list of objects that fills the layout frames. This list of objects must store the objects in the same order in which the frames appear. You can find the frame order by accessing a frame's property dialog box. In the title bar of the property dialog box a number indicates the order of a frame within the graphics list.

Printing the Layout

Sending a layout to a printer is a simple task. The following code line prints a layout on the default printer.

```
myLayout.Print
```

The Printer class represents a hardware-independent output device. Avenue statements can be used for printer setup, or to obtain setup information. The following code lines show how to access and configure the default printer.

```
' The "The" request to the
' printer class retrieves the current
' printer object.
aPrinter = Printer.The
' Open the setup dialog box.
aPrinter.SetUp
' Check to see if the printer is on line.
if (aPrinter.IsReady.Not) then
   MsgBox.Warning ("Printer off line.", "")
end
' Print to a file.
aPrinter.SetFileName (aFileName)
' Stop printing to a file.
aPrinter.SetFileName ("")
```

Application Installation

Users can access your application either by opening a predefined project file or by loading it through default settings. The project file of a frequently used application should be configured as the default project for Arc-View. However, if many applications are frequently used, each application should be maintained in a separate project file, and a menu system established in the default project file to reach each application.

Beginning with version 3 of ArcView, you can create and distribute your applications as extensions. Extensions require more work, but offer additional flexibility. Regardless of the approach you take, you need to safeguard your application by protecting its scripts. This chapter discusses protecting scripts, how to make an application the default project file, and how to create extensions.

Protecting Your Scripts

While users may be allowed to see or copy scripts, you should prevent them from modifying the scripts. Among other things, users may inadvertently crash the entire application by changing a script.

ArcView provides two levels of protection: embedding and encryption. Before undertaking either level of protection, verify that the script is compiled and free of errors.

Embedding means that the script is part of the project file, but is not displayed on the project window. Meanwhile, users can copy the script from the Script Manager dialog box. A script can be embedded through ArcView menu options or Avenue statements. When you work with multiple scripts, setting up Avenue codes for embedding is more efficient. In the following code segment, a script is displayed on the project window, embedded and removed from the project window, and then unembedded and returned to the project window.

 12-01.AVE

```
' Get the script object.
mySEd = av.GetProject.FindDoc ("AppScript")
myScript = mySEd.GetScript
if (nil = myScript) then
   MsgBox.Error ("Script is not compiled.", "")
   exit
end
' Embed the script by adding it to the
' project, and then removing
' the script editor document.
myEmbScript = myScript.Clone
av.GetProject.AddScript (myEmbScript)
av.GetProject.RemoveDoc (mySEd)
MsgBox.Info ("AppScript was embedded","")
'
' Return the embedded script to the
' project window.
myScript = av.FindScript ("AppScript")
' Create a script editor document.
mySEd = SEd.Make
mySEd.SetName ("AppScript")
' Load the script into the editor.
mySEd.SetSource (myScript.AsString)
' Unembed the script.
av.GetProject.RemoveScript ("AppScript")
MsgBox.Info ("AppScript was un-embedded.","")
```

Encryption is an irreversible process that encrypts the contents of a script. The script is listed in the Script Manager dialog box, but its contents cannot be copied or viewed. Once a script is encrypted, it cannot be decrypted. Therefore, you should always make a copy of the source code for yourself prior to encryption.

Scripts can be encrypted either separately or simultaneously. The following code segment shows how to encrypt a script. This process creates a new script object that does not replace the source code. As shown in the following code segment, the new encrypted script must be named. The script name can then be viewed in the Script Manager dialog box.

 12-02.AVE

```
' Start with the script object.
myScript = av.FindScript ("AppScript")
' Create a new encrypted script object and
' give it a name.
myEncScript = EncryptedScript.MakeFromScript
(myScript)
myEncScript.SetName
(myScript.GetName+".ENC")
' Embed it into the project file.
av.GetProject.AddScript (myEncScript)
' Save the project file.
av.GetProject.Save
```

To simultaneously encrypt all scripts, make the EncryptScripts request to the project object. The request encrypts the source for all scripts and then removes them from the project window. This process is irreversible, and source codes are lost. Consequently, you should copy the entire project or the scripts prior to encryption. Use this request with caution. In the following code segment, the project is saved and then encrypted as a new project.

```
thisProject = av.GetProject
' Save the project with source codes.
thisProject.Save
' Get a new name for the project with
' encrypted codes.
encFileName = FileDialog.Put
```

```
("","*.apr","Save Encrypted Project As")
if (nil <> encFileName) then
   thisProject.SetFileName (encFileName)
   thisProject.EncryptScripts
   thisProject.Save
   MsgBox.Info ("Encrypted project saved as"++
   encFileName.GetBaseName,"")
end
```

The EncryptScripts request encrypts all project scripts except those beginning with a period, such as *.myScript* or *.Script1*. Therefore, prior to making the request, place a period at the beginning of the script file names that you do not wish to encrypt. However, such scripts are removed from the project.

Distributing Objects

Objects can be saved to disk files. This feature allows you to distribute updates for your application, or to distribute value added objects outside a project file. ArcView stores objects in object database files. The Odb class in Avenue, which represents an object database, provides a file-based storage and retrieval system for ArcView objects. An ArcView project and its extensions are Odb files. You also create Odb files when you save legend or layout templates in a file. In the following code segment, a script is stored in an object database.

```
' Ask user to provide a new file name.
aFileName = FileDialog.Put
("script.odb".AsFileName,"*.odb","Save Object As")
if (aFileName = nil) then
   exit
end
' Create the file and Odb object.
```

```
myOdb = Odb.Make (aFileName)
' Add the script to the object database.
myScript = av.FindScript ("AppScript")
myOdb.Add (myScript)
' Save the object database.
worked = myOdb.Commit
if (worked.Not) then
   MsgBox.Error ("Unable to add"++
   myScript.GetName++"to"++
   aFileName.GetBaseName,"")
end
```

More than one object can be added to an object database. Following all the Add requests, the Commit request is sent to the Odb object only once. The Commit request saves the objects to the database file.

To retrieve an object from the object database, you need to open the database file and then use the Get request. This process is shown in the following code segment, in which the *script.odb* file is opened and all script objects are extracted.

```
' Create an Odb object by opening an
' existing object database.
myOdb = Odb.Open ("script.odb".AsFileName)
' Create a list to hold the scripts.
scriptList = {}
' Loop through all scripts in ODB
' and add it to the list.
for each index in 0..(myOdb.Count-1)
   scriptList.Add (myOdb.Get(index))
end
MsgBox.ListAsString (scriptList,"","")
```

Single User Installation

Single user installations are easy. You can make your application the default project file, or ask the user to open it as a new project file. If the user is opening the application as a new project, consider whether users should save their work over the application's original project file. In either case you can control application use by customizing the interface. For example, if you want the user to save her work to a new file you can ask her to first perform a "Save As" operation. If the following embedded script is used as the project startup script, a Save As operation is performed when the user opens the project, and the script is then removed from the new project.

 12-03.AVE

```
' Script name: Startup
' Get a new project name from the user.
newFileName = FileDialog.Put
("".AsFileName,"*.apr","Save New Project As")
' Close the project if Cancel is selected.
if (nil = newFileName) then
   av.GetProject.Close
end
' Save the project with the new name.
thisProject = av.GetProject
thisProject.SetFileName (newFileName)
thisProject.Save
' Remove this script from the project.
thisProject.RemoveScript ("Startup")
```

To establish a system default file based on the current project, send the MakeSysDefault request to the project object. Never overwrite the system project file by copying the user project or default files to the *etc* directory. If there is a user default file, it is loaded after the system default file. To avoid the user default overwriting the system default file in a single user environment, set the project as the user default file so that it is always loaded. To establish the user default project file, rename the project file to *default.apr* and copy it to one of the directories below.

❑   %HOME% for MS-Windows 3.1 and NT

❑   ArcView application folder for Macintosh

❑ $HOME for UNIX

❑   SYS$LOGIN for Open VMS

You can also send the MakeUserDefault request to the project object.

The serial number of the ArcView software running the application can be added to a script for additional security. In the following code segment, the ArcView serial number is compared to a preassigned number. If they are not the same, the application shuts down.

```
' Get ArcView's serial number.
thisSerial = av.GetSerialNumber
' Compare it to a preassigned number.
if (thisSerial <> whatShouldBe) then
   MsgBox.Error ("This is an unauthorized copy",
   "")
   av.GetProject.Close
end
```

Network Installation

Security and data access concerns are similar in a network installation of the application, but more complicated than in a single user context. First and foremost, a detailed understanding of the network and the ArcView configuration is necessary.

ArcView uses networking software to access remote data. Although remote data is handled in the same fashion as local data, the drive name assigned to the remote disk must be an input into the application. Because ArcView does not accommodate every version of every networking software package, you should consult ArcView installation documentation and the network manager prior to network installation.

ArcView may be installed centrally for multiple users to share, or on individual workstations. In either setup, the application's project file can be placed in a central location or distributed among workstations. Keep in mind that application maintenance is more difficult when the program is distributed to individual workstations, rather than maintained in a central location.

If a user customizes the interface controls of a project that is installed centrally on the network, it affects all users. You can disable the customization by placing the following code line in the project's start-up script.

```
av.SetCustomizable (false)
```

Because the above request applies to the ArcView instance, you should reset it in the project's shut-down script. When you turn off customization, you cannot access the Customize dialog box or script editor.

Therefore, use this setting with caution. Consider creating an interface control that resets this setting during your development or testing process.

Creating Extensions

Extensions offer another method to make your application available to others. The benefit of extensions is that users can load them into a project when they are needed, and unload them when they are unnecessary. Creating an extension file involves the following steps.

1. Fully develop and test your application within an ArcView project.

2. Write the following scripts for your extension.

a) *Install Script*. Loads the application into the user's project.

b) *Uninstall Script*. Removes the application from the user's project.

c) *Build Script*. Creates the extension file.

3. Build and test your extension.

While developing and testing the application, you must verify that the project file contains all objects required by the extension. Any new interface controls, data, scripts, or predefined ArcView documents required by the extension should be included in the project. The build script stores these objects in the extension file, which is an object database.

In the following sections an extension for a locator view is created. The locator view application has a pushbutton for view documents that gives a bird's-eye view of the active view. The pushbutton and the application scripts are the objects that will be stored in the extension file.

Locator Application

With this application, users can click on a button to either create or open an existing locator view. The locator view displays the full extent of the primary view and draws a rectangle showing its display extent. The locator view is automatically updated every time the extent of the primary view is changed.

Start by creating and compiling the following four scripts. Then use the customize dialog box to create a new button for the view documents. Select an icon, and associate the Locator.Show and Locator.Update scripts to the button's click and update events.

✓ *TIP: Use a common prefix for all scripts of an application for better organization.*

The Locator.Show script first checks for the existence of the locator view. If not found, it calls up another script to create a locator view. The existing locator views are tagged to their primary views. Subsequently, this script synchronizes and shows the locator view. Associate this script to the click property of the new button.

12-04.AVE

```
' Locator.Show
' Check if a locator view has already
' been created for the primary view.
pView = av.GetActiveDoc
lView = pView.GetObjectTag
if (lView = nil) then
    ' Create a new locator map.
    lView = av.run("Locator.Create",pView)
    if (lView = nil) then
      exit
    end
end
' Activate the locator view and synchronize
' its display.
lView.GetWin.Open
lView.GetWin.Activate
' Synch script needs the primary and
' locator views.
viewList ={pView,lView}
av.run("Locator.Synch",viewList)
```

Locator.Update is associated with the update property of the new button. Therefore, every time the extent of the primary view is changed, this script is executed. If the primary view has a locator view, Locator.Update calls up another script to synchronize the two views.

12-05.AVE

```
' Locator.Update
' Update the locator map.
pView = av.GetActiveDoc
lView = pView.GetObjectTag
if (lView = nil) then
    exit
end
viewList ={pView,lView}
av.run("Locator.Synch",viewList)
```

Locator.Create is called from Locator.Show to create a locator view. Tag it to the primary view, and return its object. In this script the primary view is SELF. If there is more than one theme in the primary view, the script asks the user to select one for the locator view.

12-06.AVE

```
' Locator.Create
' Create a locator map, tag it
' to its primary view, and return
' its object.
'
' SELF is the primary view.
if (SELF.Is(View).Not) then
   return (nil)
end
' Select a theme from the primary
' view to add to the locator.
themeList = SELF.GetThemes
if (themeList.Count = 0) then
   MsgBox.Error ("Cannot create a locator"++
   "for views without a theme.","")
   return (nil)
elseif (themeList.Count = 1) then
   aTheme = themeList.Get(0)
else
   aTheme = MsgBox.List (themeList,
   "Select a theme for the locator view.","")
   if (aTheme = nil) then
      return (nil)
   end
end
lTheme = aTheme.Clone
lTheme.SetVisible (true)
' Create the locator view.
lView = View.Make
lView.SetName (SELF.GetName+".Locator")
lView.SetTOCWidth (0)
lView.AddTheme (lTheme)
```

```
' Tag it to the primary view.
SELF.SetObjectTag (lView)
return (lView)
```

The last script is Locator.Synch, which is called up by other scripts to synchronize a locator view with its primary view. The SELF object in this script is a list that holds the objects of primary and locator views. Once synchronized, the locator view shows a rectangle depicting the display extent of the primary view.

 12-07.AVE

```
' Locator.Synch
' Synchronize the locator view with the
' primary view. SELF is a list of
' these views.
pView = SELF.Get(0)
lView = SELF.Get(1)
if ((pView.Is(View).Not) or
(lView.Is(View).Not)) then
    return (nil)
end
' Clear all graphics from the locator view.
lView.GetGraphics.Empty
' Confirm that the locator is at its full extent.
lView.GetDisplay.SetExtent
(lView.ReturnExtent.Scale(1.1))
' Add a rectangle to the locator view
' equal to the display extent of the primary view.
pViewExtent = pView.GetDisplay.ReturnVisExtent
lRect = GraphicShape.Make (pViewExtent)
lView.GetGraphics.Add(lRect)
```

Fully test the new application before creating the extension.

Install and Uninstall Scripts

ArcView executes the install and uninstall scripts of an extension when it is loaded into or unloaded from the current project. The purpose of these scripts is to insert or remove your application's objects in the current project. Your application scripts, however, remain inside the extension. ArcView looks inside the loaded extension files when it cannot find a script to execute.

The Locator.Install script, shown below, places the pushbutton from the extension file onto the button bar of view documents. The objects are placed in the extension file by the build script in the last step. If you have more than one object to insert, document the order in which they are to be placed in the extension file. Because the extension file is an object database, the install program can access the objects only by their placement index. For example, the first object placed in the extension file by the build script can be accessed by the install script as object zero, the second object as one, and so on.

 12-08.AVE

```
' Locator.Install
' Move the button from the extension
' file into the current project.
' SELF is the extension file.
thisProject = av.GetProject
if (thisProject = nil) then
    return (nil)
end
' Add the button to the end of the
' button bar for view documents.
theBBar = thisProject.FindGUI("View").GetButtonBar
theBBar.Add (SELF.Get(0),999)
```

The uninstall script, named Locator.Uninstall, removes the button from the button bar.

 12-09.AVE

```
' Locator.Uninstall
' Remove the button from
' the current project.
' SELF is the extension file.
thisProject = av.GetProject
if (thisProject = nil) then
   return (nil)
end
' Do not uninstall when project is closing.
if (thisProject.IsClosing) then
   return (nil)
end
' Remove the button.
theBBar = thisProject.FindGUI("View").GetButtonBar
theBBar.Remove (SELF.Get(0))
```

You must store the preceding install and uninstall scripts in the application project along with the other locator scripts.

Build Script

Once you have all the objects required by your application, and the install and uninstall scripts are in the ArcView project, you are ready to add the build script. This script creates the extension file and adds the application objects to it.

CODE 12-10.AVE

```
' Locator.ExtBuild
' Create the locator extension.
' The install and uninstall scripts
' come first. Verify that they are compiled.
theProject = av.GetProject
theInstallScript =
theProject.FindDoc("Locator.Install").GetScript
theUninstallScript =
theProject.FindDoc("Locator.Uninstall").GetScript
if ((theInstallScript = nil)
or (theUninstallScript = nil)) then
    MsgBox.Error ("Install or Uninstall script"++
    "is not compiled.","")
    exit
end
' Create the extension file with .avx suffix.
locatorExt = Extension.Make ("locator.avx".AsFileName,
"Locator",theInstallScript,theUninstallScript,{})
' Add the pushbutton and required scripts.
theBBar = theProject.FindGUI("View").GetButtonBar
locatorExt.Add (theBBar.FindByScript("Locator.Show"))
locatorExt.Add
(theProject.FindDoc("Locator.Create").GetScript)
locatorExt.Add
(theProject.FindDoc("Locator.Show").GetScript)
locatorExt.Add
(theProject.FindDoc("Locator.Synch").GetScript)
locatorExt.Add
(theProject.FindDoc("Locator.Update").GetScript)
' Add extension description.
locatorExt.SetAbout ("Locator")
' Save the changes to the extension file.
locatorExt.Commit
```

Execute the preceding script to build your extension file. Depending on the system, move the extension file to the *ext16* or *ext32* directory of ArcView's install directory. ArcView also looks for extensions in the directory defined by the environment variable USER-EXT. Test the extension by starting a new project and loading the extension. If you encounter problems, or decide upon an enhancement, you must return to the original application project, make the changes and re-execute the build script.

The Make request creates an object of the Extension class and requires five parameters. The first parameter is the file name for the extension file. ArcView over-writes the file if it already exists. The second parameter is the name of the extension followed by the install and uninstall scripts. The last parameter is a list of the other extensions on which this one is dependent. Arc-View loads the referenced extensions before loading this one. You can use this feature to create shared extension libraries or break up a large application into several smaller extensions. You can send the GetExtensions request to the Extension class to access and place extension objects in this dependency list.

Other Scripts

The Extension class offers three other event-driven scripts that can give you more control over using an extension.

Load Script

You can define a load script by sending the SetLoad-Script to the extension object. The load script is executed only once, when the extension file is opened and loaded into the project. Use this script to perform one-time initialization.

Unload Script

Specify the unload script with the SetUnloadScript request. This script is executed only once, when the extension is removed from the project. Use it for final cleanup.

CanUnload Script

ArcView executes this script just prior to unloading an extension. If the script returns true, then ArcView continues removing the extension. If it returns false, ArcView cancels the removal process. Assign this script by sending the SetCanUnloadScript to an extension object.

Address Matching

GIS is a tool for managing and analyzing data with spatial attributes, that is, latitude and longitude, or northing and easting. For most people, however, street addresses are the most common spatial attribute, not X and Y. Real estate taxes are assessed by property address, crime is reported by city block, and new commercial sites are selected along major roadways.

An address specifies location in the same manner as geographic coordinates. But addresses must be converted to geographic coordinates before ArcView or any other GIS software can display and use them. The process of transforming an address (text string) into geographical coordinates is called address matching, or "geocoding."

Geocoding with ArcView

ArcView provides the tools for geocoding either through Avenue or directly through its interface. Because address matching is a rather complicated process, you should familiarize yourself with it using ArcView's interface. Briefly, address matching with ArcView involves the following steps.

1. Make a theme matchable.

2. Add address events.

3. Create a geocoded theme.

4. Process unmatched events.

Matched locations are depicted by graphical symbols. The address matching steps are described below.

Making a Theme Matchable

For address matching, you need a theme with specific searchable attributes. Each type of address format—linear, polygon and point—has different address components.

A linear address format requires a theme with line features, such as a street network. Typically, the address component attributes for this format are left-from, left-to, right-from, right-to, street-name, and type. For example, an arc within a coverage could have the values listed below.

❏ left-from: 1301

❏ left-to: 1399

❏ right-from: 1300

❏ right-to: 1398

❏ street-name: WISCONSIN

❏ type: AV

The from-to ranges indicate the possible numbers that could fall within a specific block, and the left and right sides divide the numbers into odd and even groups. The preceding example showed values for a block along Wisconsin Avenue. Address numbers along the block range from 1300 to 1399, with odd numbers on the left and even numbers on the right. Additional optional fields, such as direction, can be used to further define an address.

The linear format, used most frequently for locating addresses within urban areas, can also contain zone information. Zone refers to an additional attribute on each side of the line segment used to discriminate among streets with similar names. For example, a zip code or city name can serve as zone information. Thus, upon searching for Main Street within a specific county, a city name can narrow the search scope. Examples of ZIP codes used as zone attributes follow: left-zone: 20016, and right-zone: 20017.

If an entire street segment is inside a ZIP code area, the left and right values in the above example would be the same. Attribute names defining address components can be different from the fields mentioned here. In the process of making a theme matchable, ArcView searches for default attribute or field names (left-from, l-f, l_f, street-name, st-name, st-n, etc.). However, if ArcView does not find what the program identifies as a particular attribute name, or selects the wrong attribute for what the user has in mind, the field names that define the address components can be set or changed.

The polygon address format requires themes with polygon features. In this format, a single attribute can be assigned as the address component, such as census tract or ZIP code area. In this case, the value of the census tract number or ZIP code is used in search and match routines.

The point address format requires themes with point features, and is useful in locating items with identifying names or numbers. Similar to the polygon format, only a single attribute is necessary. For instance, in a theme showing sewer system valves, the single attribute could be valve number. Another example is a theme comprised of landmarks where the attribute is building name.

You can make ARC/INFO coverages and ArcView shape files matchable. ArcView creates a match source file and a geocode index file for matchable themes. Thereafter the theme remains matchable when loaded into any project.

Adding Address Events

Events are tabular data that can be added to views as themes. Initially, events are added to the project in the form of tables that contain geographic information, such as street addresses or building names. Events on a base theme can be located by using this geographic information. The base theme contains the address components described in the previous section.

An example of an address event is students' home street addresses in a particular school district. The base theme for this address event could be a street map of the school district. Another example of an address event is the number of home mortgages by census tract, where the base theme is a map of several contiguous census tracts.

Creating a Geocoded Theme

Once a base theme is matchable and an appropriate events table is added to the project, ArcView attempts to locate each event on the base theme. In this process, ArcView creates a shape file theme that contains a symbol at the location of each matched event. The geocoded theme contains point features for both matched and unmatched events. The features for unmatched events are placed in the theme as null point shapes.

Processing Unmatched Events

As mentioned above, the geocoded theme created by ArcView contains both matched and unmatched events as point features. The unmatched events are represented by null point shapes in the theme. Any event in the geocoded theme can be rematched. The two approaches for processing unmatched events are manual editing to correct errors in events, and relaxation of the matching requirements to increase the probability of finding a match.

Geocoding with Avenue

Avenue can be used to install automatic address matching tasks in an application. This approach is useful when the same type of event data must be repeatedly matched to the same base theme. For example, if a monthly crime report includes a map of all incidents, you can create a script to match the crime database against your base theme. The following sections describe how to make a theme matchable and how to create an address events theme.

Avenue can also be used to add event tables. Adding event tables is similar to adding any attribute table to a project. (See Chapters 8 and 9 on adding attribute tables.)

Processing unmatched events is generally an on-line manual task because it involves reviewing why an event failed to match, and perhaps correcting data entry errors. There is little value in using Avenue for processing unmatched events unless you decide to rematch everything after relaxing match requirements.

The steps required to make a theme matchable and to create a geocoded theme are shown in the following two scripts. In the first script a theme called *Street* in the *County* view is made matchable for the linear address format. In the second script, an event table with addresses for bank locations is matched against the Street theme. The steps are explained in detail following the scripts.

13-01.AVE

```
' Prepare for address matching by making
' a theme matchable.
theProject = av.GetProject
theTheme = theProject.
FindDoc("County").FindTheme("Street")
if (nil = theTheme) then
   MsgBox.Error (
   "Unable to access Street theme in County view",
   "")
   exit
end
'
' Verify whether the theme is already matchable.
if (theTheme.isMatchable) then
   MsgBox.Warning ("Street is already matchable",
   "")
   exit
end
'
' The GetDefStylesODB request to the AddressStyle
' class returns a file name object corresponding
' to the style object database supplied with
' ArcView.
addrStyleFilename = AddressStyle.GetDefStylesODB
```

```
if (nil = addrStyleFilename) then
   MsgBox.Error (
   "Unable to find the default address style ODB",
   "")
   exit
end
'

' The list of styles can be extracted from the
' address style file name, and the desired style
' can be found within that list.
addrStyleList = AddressStyle.GetStyles
(addrStyleFilename)
addrStyle = AddressStyle.FindStyle
("US Streets with Zone")
'

' Associate the known fields to
' the address components.
theVTab = theTheme.GetFTab
attList = {}
'

' Setup for the US Streets with Zone style.
nameList = {"LADD_FROM", "RADD_RIGHT",
"LADD_TO", "RADD_TO", "NONE", "NONE",
"STREET", "TYPE", "NONE", "LZIP", "RZIP"}
for each fldName in nameList
   if (fldName = "NONE") then
      ' NONE indicates that there is no field to match.
      attList.Add ("") ' Add a null value.
      continue
   end
   ' Get the field object.
   aField = theVTab.FindField (fldName)
   if (nil = aField) then
      MsgBox.Error ("Unable to access required field"++
      fldName,"")
      exit
   end
   attList.Add (aField)
end
```

```
'
' Create a MatchSource object.
aMatchSource = MatchSource.Make (addrStyle,
theTheme, attList)
'
' Now assign the MatchSource object to the theme.
theTheme.SetMatchSource (aMatchSource)
'
' Verify whether the theme is matchable.
if (theTheme.IsMatchable) then
    MsgBox.Info (theTheme.GetName++"is matchable",
    "")
else
    MsgBox.Warning (theTheme.GetName++
    "Is not matchable",
    "")
end
```

CODE 13-02.AVE

```
' Perform address matching.
'
' Get the MatchSource object.
theView = av.GetProject.FindDoc("County")
theTheme = theView.FindTheme("Street")
if (theTheme.IsMatchable) then
    theMatchSource = theTheme.GetMatchSource
else
    MsgBox.Warning (theTheme.GetName++
    "Is not matchable",
    "")
    exit
end
'
' Get the event table VTab object.
theVTab = av.getProject.FindDoc("Banks").GetVTab
'
' Verify that MatchSource uses the zone,
```

```
' then get zone and address fields
' from the event table.
if (theMatchSource.HasZone) then
   zoneField = theVTab.FindField ("ZIP")
end
addrField = theVTab.FindField ("ADDRESS")
'
' Create GeoName object.
theGeoName = GeoName.Make (theMatchSource,
theVTab, addrField, zoneField)
'
' Set a shape file name to store the new
' geocoded theme.
theGeoName.SetOutFileName ("BANKS.SHP".AsFileName)
'
' Initialize the geocoded theme.
geoVTab = theMatchSource.InitGeoTheme (theGeoName)
'
' Use US_ADDR address standardization
' file for US Streets style and US Streets
' with Zone style.
aMatchKey = MatchKey.Make ("us_addr.stn")
if (nil = aMatchKey) then
   MsgBox.Error ("Unable to create a match key",
   "")
   exit
end
'
' Create a MatchCase object.
aMatchCase = MatchCase.Make
(theMatchSource, aMatchKey)
'
' Create default matching preference values.
aMatchPref = MatchPref.Make
'
```

```
' Get number of records to match.
numRecs = theVTab.GetNumRecords
'
' Start the loop to go through all records.
for each recNum in theVTab.GetDefBitMap
   ' Set the progress bar.
   av.SetStatus (((recNum+1)/numRecs)*100)
   ' Standardize the address string.
   aMatchKey.SetKey (theVTab.ReturnValueString
   (addrField,recNum))
   ' Search for candidates; search request returns
   ' number of candidates.
   numCand = theMatchSource.search (aMatchKey,
   aMatchPref.GetPrefVal(#MATCHPREF_SPELLWEIGHT),
   aMatchCase)
   if (numCand = 0) then
      theMatchSource.WriteUnMatch (recNum, aMatchKey)
   else
      ' Score and get the best candidate.
      aMatchCase.ScoreCandidates
      cand = aMatchCase.GetBestCand
      ' Write it to the GeoName data source.
      TheMatchSource.WriteMatch
      (recNum, aMatchKey, cand)
   end
end
'
' End the matching process; Avenue closes index file.
theMatchSource.EndMatch
'
' Create and add the address event theme.
geoTheme = Theme.Make (theGeoName)
theView.AddTheme (geoTheme)
```

Object Model for Address Matching

Reviewing the object model shown in the following figure can help in understanding the geocoding process.

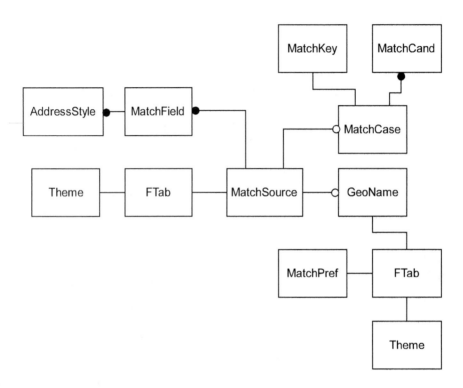

Object model for address matching.

MatchSource is at the heart of the geocoding process. A MatchSource object is created on an address style and a base theme with address ranges. Various address styles use different sets of match fields, which can be required or optional. If a match field is required, a corresponding field in the base theme table must exist. A theme becomes matchable when its MatchSource property is set.

The process of address matching begins by establishing an instance of MatchKey that parses addresses. Searching the base theme for a match can create the MatchCase object, which is a list of candidate matches. Each candidate is scored based on how well it has matched.

The result of address matching is a new theme with symbols at matched areas. An instance of the GeoName class maintains the information needed to create a geocoded theme. The criteria to accept or reject a match candidate is stored in an instance of the MatchPref class.

Making a Theme Matchable

A MatchSource object is created in order to make a theme matchable with Avenue. The Make request for the MatchSource class requires a theme, an address style, and a list of fields in the theme's attribute table that corresponds with the fields required by the address style. Steps for this task follow.

1. Select a theme.

2. Select an address style.

3. Associate attribute fields.

4. Create a MatchSource object.

Selecting a Theme

As shown in the following code segment, a theme is selected by accessing its object. When a MatchSource object called *aMatchSource* is created later in the chapter, *theTheme* becomes matchable by setting its MatchSource attribute.

```
theProject = av.GetProject
theTheme = theProject.FindDoc
("County").FindTheme("Street")
if (nil = theTheme) then
   MsgBox.Error
   ("Unable to access Street theme in County view",
   "")
   exit
end
'
' Check to see if theme is already matchable.
if (theTheme.isMatchable) then
   MsgBox.Warning ("Street is already matchable",
   "")
end
```

Selecting an Address Style

The address style defines the address format and its required and optional fields. ArcView's address styles appear below.

❏ U.S. Streets with Zone

❏ U.S. Streets

❏ U.S. Single Range with Zone

❏ U.S. Single Range

❏ U.S. Single House with Zone

❏ U.S. Single House

❏ Single Field

❏ Zip+4

❏ Zip+4 Range

Styles are stored in an object database in the *geocode* directory; the file name for this database is *addstyle.db*. Each style has a different set of required and optional fields.

Creating an address style object is a two-step process.

1. Access the object database containing the address styles.

2. Search for the desired address style.

An AddressStyle class object is created when an address style is selected. In the following code segment, the default style database supplied with ArcView is accessed, and the *US Streets with Zone* style is selected.

```
' GetDefStylesODB request to the AddressStyle class
' returns a file name object corresponding to the
' style object database supplied with ArcView.
addrStyleFilename = AddressStyle.GetDefStylesODB
if (nil = addrStyleFilename) then
```

```
    MsgBox.Error
    ("Unable to find the default address style ODB",
    "")
    exit
end
'
' The list of styles can be extracted from the
' address style file name and the desired style
' can be found in the list.
addrStyleList = AddressStyle.GetStyles (addrStyleFilename)
addrStyle = AddressStyle.FindStyle
("US Streets with Zone")
```

Each style has a set of default parameters that can be reviewed and altered through Avenue requests. For example, when an event is located, a symbol is placed at a predefined distance from the street. This distance, measured in map units, is called the "offset value." As shown in the following statements, the default offset value for an address style can be changed.

```
' Change offset value and save it to a
' new style object database for future use.
addrStyle.SetDefOffset (newValue)
AddressStyle.SaveStyles ("mystyle.db".AsFileName)
```

Associating Attribute Fields

An address style must be selected prior to associating fields from the event table. An address style maintains the required and optional address components in a predetermined order. To associate attribute fields from the event table to address components, an event table's list of fields must be created in the same order. If optional address components do not exist, or if you do not wish to provide corresponding attribute fields, nil objects corresponding to these components are stored in their respective places on the list.

Creating the fields list can be tricky. Unless the field names are known ahead of time, the user may be requested to select the fields. For example, if the same event table is matched every month, you could code the field names directly into your script. However, if the event table and its fields are not always the same, you should display the fields list and ask the user to determine how the fields correspond to the address components. Both approaches are presented in the following two code segments. In the first segment, the field names are known ahead of time. In the second, a complete fields list is presented to the user for selecting each component.

```
' Associate the known fields to the address
' components.
theVTab = theTheme.GetFTab
attList = {}
'
' Set up for the US Streets with Zone style.
nameList = {"LADD_FROM", "RADD_RIGHT",
"LADD_TO", "RADD_TO", "NONE", "NONE",
"STREET", "TYPE", "NONE", "LZIP", "RZIP"}
for each fldName in nameList
if (fldName = "NONE") then
   ' NONE indicates that there is no field to match.
   attList.Add ("") ' Add a null value.
   continue
end
' Get the field object.
aField = theVTab.FindField (fldName)
if (nil = aField) then
   MsgBox.Error ("Unable to access required field"++
   fldName,"")
```

```
        exit
    end
    attList.Add (aField)
end
```

CODE

13-3.AVE

```
' Associate the event table fields with
' address components by presenting them
' to the user. Store the results in attList.
theTheme = av.GetProject.FindDoc("County")
FindTheme("Street")
theVTab = theTheme.GetFTab
attList = {}
'
' The GetDefStylesODB request to the AddressStyle
' class returns a file name object corresponding
' to the style object database supplied with
' ArcView.
addrStyleFilename = AddressStyleGetDefStylesODB
if (nil = addrStyleFilename) then
    MsgBox.Error(
    "Unable to find the default address style ODB,"
    "")
    exit
end
' The list of styles can be extracted from the
' address style file name, and the desired style
' can be found within that list.
addrStyleList = AddressStyleGetStyles
(addrStyleFilename)
addrStyle = AddressStyle.FindStyle
("US Streets with Zone")
'
' Get the list of address components.
matchFieldList = addrStyle.GetMatchFields
'
' As each address component is presented
' to the user, identify it as required or optional.
for each fld in MatchFieldList
```

```
        if (fld.IsRequired) then
          optionality = "Required"
        else
          optionality = "Optional"
        end
      while (True) ' Stay in loop until you get what you want.
        aField = MsgBox.List (theVTab.GetFieldList,
        fld.GetName++optionality,"")
      ' aField is nil if user clicked on the Cancel button.
      if (nil = aField AND (optionality = "Required")) then
        ' Null values cannot be allowed for required components.
          stopIt=MsgBox.YesNo
          ("Do you want to cancel address matching?",
          "", False)
          if (stopIt) then
            exit
          else
            ' Try again to get the field name.
            continue
          end
        end
        break
      end
      ' Add field name to the list of attribute fields.
      attList.Add (aField)
end
```

Creating a MatchSource Object

A MatchSource object is a feature source that accepts
matching requests. When a MatchSource is created, a
theme is made matchable. Creating a MatchSource
object is easy once the theme, address style, and
fields list are selected or prepared. These objects are
simply supplied as parameters of the Make request to
the MatchSource class. See the following code segment for the format of the Make request.

```
' Create a MatchSource object.
aMatchSource = MatchSource.Make (addrStyle,
theTheme, attList)
'
' Now assign the MatchSource object to the theme.
theTheme.SetMatchSource (aMatchSource)
'
' Check to verify that the theme is matchable.
if (theTheme.IsMatchable) then
  MsgBox.Info (theTheme.GetName++"is matchable",
  "")
else
  MsgBox.Warning (theTheme.GetName++
  "is not matchable","")
end
```

Creating a Geocoded Theme

The process of address matching results in a new theme that displays a symbol for each matched event. The following steps are required for address matching.

1. Retrieve the match source.

2. Select the event table.

3. Initialize a GeoName.

4. Match addresses.

5. Create a geocoded theme.

Retrieving the Match Source

A matchable theme has a MatchSource object. This object is necessary in order to search for each address event. In the following script, a theme is tested to determine whether it is matchable, and then the MatchSource object is retrieved.

```
' Get the MatchSource object.
theTheme = av.GetProject.FindDoc("County")
FindTheme("Street")
if (theTheme.IsMatchable) then
   theMatchSource = theTheme.GetMatchSource
else
   MsgBox.Warning (theTheme.GetName++"Is not matchable",
   "")
end
```

Selecting the Event Table

The event table holds the field or fields that describe the address. First, the event table and the fields to match must be accessed. Given the use of the US Streets with Zone address style, two fields from the event table are required. One field holds the address while the other field contains the zone.

The contents of the zone field or the address field for the polygon or point address format are simple. In these cases, a single value is matched against another single value in the matchable theme. The address field for the linear address style requires a more complicated procedure because the field is composed of several components, and each is matched against a different field in the matchable theme. Because the address field in the event table is a single text field, ArcView must follow a set of rules to identify each address component in the field. For example, ArcView "assumes" that the first numerical value is the house number, and that a single alphanumeric (e.g., *W*) before or after the street name indicates direction (west). Standards for identifying address components are found in the ArcView documentation.

The following code segment shows how to create address and zone field objects.

```
' Retrieve the VTab object for the event table.
theView = av.GetProject.FindDoc("County")
theTheme = theView.FindTheme ("Street")
theMatchSource = theTheme.GetMatchSource
theVTab = theEventTable.GetVTab
'
' Verify that MatchSource uses zone,
' then retrieve zone and address fields
' from the event table.
if (theMatchSource.HasZone) then
   zoneField = theVTab.FindField ("ZIP")
end
addrField = theVTab.FindField ("ADDRESS")
```

Initializing a GeoName

A geocoded theme is a feature source that holds a record for each record in the event table. Initially, a geocoded theme has a null shape for each event and its match status is set to *unmatch*. The new theme is geocoded with the Search or Rematch requests, as discussed in the next section.

A GeoName object contains the information required to create a geocoded data source. This object is created from the MatchSource, event table, address field, and zone field. In addition, an output file name must be assigned to store the geocoded data set in a shape file format. In the following code segment, a GeoName object is created and used to initialize a geocoded theme.

```
' Create GeoName object.
theGeoName = GeoName.Make (theMatchSource,
theVTab, addrField, zoneField)
'
' Set a shape file name to store the new
' geocoded theme.
theGeoName.SetOutFileName ("BANKS.SHP".AsFileName)
'
' Initialize the geocoded theme.
theMatchSource.InitGeoTheme (theGeoName)
```

Matching Addresses

Search and WriteMatch requests to the MatchSource object are used to find a match and write it to the geocode theme. Before using these requests, however, three objects from the *MatchKey*, *MatchCase*, and *MatchPref* classes must be created.

The MatchKey class standardizes the address field of the event table. A MatchKey object is created from the information supplied in standardization files. ArcView provides standardization files for each address style. These files have an *stn* extension and reside in the *geocode* directory. The standardization files are related to several other files, which define expected address components, possible and allowed values, and a scoring system. For example, for the U.S. Streets address style defined in the *us_addr.stn* file, Avenue searches for the components listed below in an address field.

❑   House number

❑   Pre-direction

❑   Pre-type

❑   Street name

❑   Type

❑   Suffix direction

While a "hit" for every component is not expected, the higher the number of matched components, the higher the match score. A match score is computed to assist in selecting the best among multiple matches. The following code segment creates a MatchKey object for the U.S. Streets address style.

```
' Use US_ADDR address standardization
' file for US Streets style and US Streets
' with Zone style.
aMatchKey = MatchKey.Make ("us_addr.stn")
if (nil = aMatchKey) then
   MsgBox.Error ("Unable to create a match key",
   "")
   exit
end
```

A MatchCase object is comprised of a list of matching candidate records for a single event. The candidates can be scored to select the best match. In the following code line, a MatchCase object is created. This object is populated every time a Search request is sent to the MatchSource object.

```
' Create a MatchCase object.
aMatchCase = MatchCase.Make (theMatchSource,
aMatchKey)
```

A MatchPref class object contains geocoding preferences such as the minimum match score. Default preference values are stored in the *mprefdef.db* file in the *geocode* directory. Default preference values are listed below. (See the ArcView documentation for a discussion of these parameters.)

❏ Spell weight: 80

❏ Minimum match score: 60

❏ Minimum candidate: 30

❏ No user review: 1

Through the MatchPref class, Avenue allows you to change the default values to match your needs. A MatchPref object is created in the code line below. Because the attributes for this object are not changed, the default values are assumed.

```
' Create default matching preference values.
aMatchPref = MatchPref.Make
```

Once the MatchKey, MatchCase, and MatchPref objects are created, address matching can begin. This process generally requires iteration through all records of the event table in an attempt to match each record. The matching process for each event involves the following tasks.

1. Standardize the address string.

2. Send a Search request to the MatchSource object.

3. If no matching candidate is found, send a Write-UnMatch request to the MatchSource object.

4. If matching candidates are found, score them, select the best candidate, and send a WriteMatch request to the MatchSource object.

The preceding tasks are demonstrated in the following code segment. A progress bar showing the proportion completed as matching occurs is included.

```
' Get number of records to match.
numRecs = theVTab.GetNumRecords
'
' Start the loop to go through all records.
for each recNum in theVTab.GetDefBitMap
    ' Set the progress bar.
   av.SetStatus (((recNum+1)/numRecs)*100)
    ' Standardize the address string.
   aMatchKey.SetKey (theVTab.ReturnValueString (addrField,recNum))
    ' Search for candidates; search request returns
    ' number of candidates.
   numCand = theMatchSource.search (aMatchKey,
   aMatchPref.GetPrefVal(#MATCHPREF_SPELLWEIGHT),
   aMatchCase)
   if (numCand = 0) then
     theMatchSource.WriteUnMatch (recNum, aMatchKey)
   else
     ' Score and retrieve the best candidate.
     aMatchCase.ScoreCandidates
     cand = aMatchCase.GetBestCand
     ' Write the best candidate to the
```

```
      ' GeoName data source.
      TheMatchSource.WriteMatch (recNum, aMatchKey, cand)
    end
end
'

' End the matching process; Avenue closes index file.
theMatchSource.EndMatch
```

Creating the Geocoded Theme

After creating the *GeoName* data source, initializing the address theme, and recording the match results, the only remaining task is to create the theme. In the following code segment, the address event theme is created and added to the view.

```
' Create and add the address event theme.
geoTheme = Theme.Make (theGeoName)
theView.AddTheme (geoTheme)
```

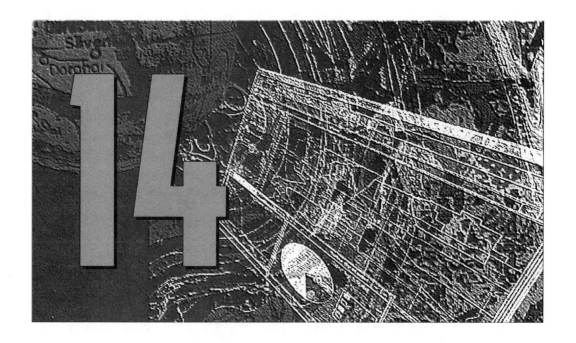

Integration

The ability to access system resources enhances an application's capabilities. Avenue allows you to access the system clipboard and environment variables, and to issue operating system commands inside your program. Avenue's integration capabilities, however, go far beyond these tasks.

Avenue supports Dynamic Data Exchange (DDE), Apple Scripts, and Remote Procedure Call (RPC). For data transfer between applications, DDE is used in the

Microsoft Windows environment, Apple Scripts for
Macintosh, and RPC in the UNIX environment. For
example, if you use a spreadsheet which supports DDE
and you have created a budget analysis model using
numbers from ArcView, the spreadsheet can be linked
to ArcView. As the numbers change in ArcView,
spreadsheet numbers are automatically updated.

This chapter covers accessing the clipboard and system
variables, issuing system commands, and the use of
DDE, Apple Scripts, and RPC. As an extension of issuing system commands, the execution of ARC/INFO
AMLs is also discussed. Topics discussed in this chapter
are generally platform dependent. Consequently,
attention to the specific class or request for your hardware and operating system platform is required.

Accessing the Clipboard

The system clipboard holds text strings and graphics,
while the application (ArcView) clipboard holds ArcView
objects. The system clipboard stores graphics in metafile
format for Windows and PICT format for Macintosh.

The Clipboard class in Avenue references the application clipboard. Although ArcView does not have a
class for the system clipboard, text strings and graphics can be exchanged with the system clipboard
through the application clipboard. The two clipboards do not automatically synchronize.

The following code line shows how the clipboard
object is accessed.

```
appClipBoard = ClipBoard.The
```

Because the clipboard object is a specialized list, you can use the Add and Get requests to copy and paste an object to or from the application clipboard. The Add request clones an object and adds the clone to the clipboard. As seen in the following code segment, the object is cloned in order to paste from the clipboard.

```
' Place the first two themes in View1
' on the clipboard.
appClipBoard = ClipBoard.The
theView = Av.GetProject.FindDoc ("View1")
appClipBoard.Add (theView.GetThemes.Get(0))
appClipBoard.Add (theView.GetThemes.Get(1))
'
' Paste the themes on the clipboard to
' the View2 document.
newView = Av.GetProject.FindDoc ("View2")
newView.AddTheme (appClipBoard.Get(0).Clone)
newView.AddTheme (appClipBoard.Get(1).Clone)
```

To remove all objects from the application clipboard, send the Empty request as shown in the code line below.

```
ClipBoard.The.Empty
```

Text and graphics can be moved between the system and application clipboards. When you send the Update request to the ClipBoard class object, ArcView checks to see if there are any objects on the application clipboard. If one or more objects are found, the contents of the application clipboard are copied to the system clipboard. In the event no objects are found in the application clipboard, ArcView then checks for a new item in the system clipboard. If there is a new item in the system clipboard it is added to the application clipboard.

The string objects in the application clipboard are converted to text blocks before copying to the system clipboard. Text in the system clipboard is converted to a string object when copied to the application clipboard. This entire transfer process occurs when the Update request is sent to the clipboard object. A Boolean object is returned by the Update request, and is set to True if the system clipboard has acquired a new item since the last update. The following code segment shows how the Update request works.

 14-01.AVE

```
' Synchronize the clipboards.
appClipBoard = ClipBoard.The
newClip = appClipBoard.Update
if (newClip) then
    ' The contents of the system clipboard are
    ' added to the application clipboard.
    MsgBox.ListAsString (appClipBoard,
    "Contents of the application clipboard",
    "")
else
    if (appClipBoard.Count > 0) then
        ' The contents of the application clipboard
        ' are copied to the system clipboard.
        MsgBox.ListAsString (appClipBoard,
        "Contents of the application clipboard",
        "")
    else
        MsgBox.Info ("Clipboards are empty", "")
    end
end
```

Accessing the Operating System

The System class provides the tool for accessing operating system variables and commands. For example, if you want to start another program or obtain the value of an environment variable, appropriate requests are sent to the System class. The System class has no objects; all requests are sent directly to the class.

Accessing Environment Variables

Environment variables are retrieved or set through the GetEnvVar and SetEnvVar requests. For example, to locate the directory that holds ArcView data, the AVDATA value can be retrieved, assuming this variable is set in the *autoexec.bat* file. The following code segment searches for AVDATA; if AVDATA is not found, the variable is set to a user-supplied value.

CODE 14-02.AVE

```
' Get the AVDATA environment variable.
avData = System.GetEnvVar ("AVDATA")
if (nil = avData) then
    ' If not set, get AVDATA value from the user, and
    ' then set it.
    avData = MsgBox.Input
    ("AVDATA variable is not set",
    "Enter AVDATA value: ","")
    if (nil = avData) then
      exit
    end
    System.SetEnvVar ("AVDATA",avData)
end
```

Additional valuable information can be obtained from the system. For example, screen resolution in pixels can be determined by sending the ReturnScreen-SizePixels request. This request returns a Point object.

The object's X value is screen width, and the Y value is screen height.

Issuing Operating System Commands

The System class can be used for direct issue of system commands. For instance, you can easily create a new directory or copy one file to another by using system commands. As shown in the following code segment, the Execute request can issue system commands. The contents of the command line (*commandLine*) are platform dependent.

```
' Make a quick backup of the project file.
projectName = Av.GetProject.GetFileName
baseName = projectName.GetBaseName
backupName = baseName.SetExtension ("BAK")
commandLine =
"Copy"++projectName.AsString++backupName
System.Execute (commandLine)
```

Executing AMLs

AMLs are executed from an Avenue script by sending the Execute request to the System class. The code below runs an AML called *getvalve*.

```
' Run AML.
System.Execute ("arc getvalve")
```

Avenue does not wait for the Execute request to complete its process. As soon as the command line is passed to the operating system, Avenue executes the next line of the script.

Implementing Dynamic Data Exchange

Dynamic Data Exchange (DDE) in Microsoft Windows is used to create a relationship between data stored in one application with data stored in another. This mechanism enables two applications to continuously and automatically exchange data. Both applications must support DDE to be able to participate in this data exchange.

In a DDE conversation, the application that initiates the conversation is known as the "client," and the responding application is called the "server." ArcView can be both client and server simultaneously.

To initiate a DDE conversation and exchange data, the client application must specify the following:

❑ Server application name

❑ Conversation topic

❑ Item

The name of the server application and its executable file are often the same. For instance, if you have created a Visual Basic program named *myvb.exe*, its server name is *myvb*. The server name of commercial applications that support DDE is included in respective documentation. An example here is the server name *WinWord,* specified in the Microsoft Word for Windows documentation.

The conversation topic is usually a data unit meaningful to the server. For example, in Microsoft Word, the topic can be a file name. A DDE conversation begins after the server recognizes the topic. Once a DDE conversation is established, the application or topic name cannot be changed. Many applications that support DDE also support a topic called "System."

Finally, "item" is the term for the piece of data to be exchanged, such as a field value in a database application.

DDE conversations are often called "hot links" or "cold links." The difference between the two link types is in how data is updated when changes occur. In a hot link, the server supplies new data to the client every time the data changes. In a cold link, the client must request the new data.

ArcView as DDE Client

Through Avenue, ArcView can initiate a DDE conversation. To begin a DDE and receive data, you need to know the server, topic, and item names. Because all three are application dependent, a review of the appropriate documentation is recommended.

For example, if the server application is a Visual Basic program, the executable file name minus its extension is the server name. The topic name is set for each Visual Basic form as its LinkTopic property and it can be any character string. The Name property of a control on the Visual Basic form becomes the item name.

Assume that you have a Visual Basic program named *myvb.exe* that contains a form. The form's LinkTopic property is set to Form1. The form has a text box control named Text1. In the following code segment, a DDE conversation is established with *myvb,* and the contents of its text box are retrieved. Keep in mind that *myvb* must be running in order to establish a DDE conversation. The program could be minimized on your screen.

```
' Establish a DDE conversation with myvb.exe.
aDDEClient = DDEClient.Make ("myvb","Form1")
if (aDDEClient.HasError) then
   MsgBox.Error (aDDEClient.GetErrorMsg, "")
   exit
end
'
' Get the text string in the Text1 text box.
textValue = aDDEClient.Request ("Text1")
```

The Make request sent to the DDEClient class requires two parameters. The first parameter is the server name and the second is the topic name. Once a DDEClient object is created, you can request information, send information, or execute a task. Several DDE conversations can be opened simultaneously to the same or different servers.

ArcView as DDE Server

The default project start-up script establishes ArcView as a DDE server. In the event the default script is changed or the DDE server is halted, the DDE server routine can be started by the following statement:

```
DDEServer.Start
```

If the above statement is executed more than once, two instances of the ArcView server are established. This situation causes errors when other applications initiate DDE conversations. To avoid executing the Start request more than once, insert a Stop request before the Start request as shown in the following code segment.

```
' Start DDE server support, but make
' sure that it is not duplicated.
DDEServer.Stop
DDEServer.Start
```

The ArcView server name is *ARCVIEW*, and the ArcView server supports only one topic called *SYSTEM*. Items in the ArcView server can be any string type object. For example, if you are developing a Visual Basic program to show the name of the ArcView project in a text box, the text box must have the following properties:

```
LinkTopic = ARCVIEW|SYSTEM
LinkItem = av.GetProject.AsString
```

Implementing Remote Procedure Call

The Remote Procedure Call (RPC) is a logical client/server communication system that supports network applications in a UNIX environment. A client process makes an RPC and waits for a response to be returned from the server process. Developers of distributed applications avoid the details of the network interface with the use of RPC. The following information items are required to create an RPC:

❑ *Server machine name* is the host name for the RPC server.

❑ *Server identification number* is the number that identifies the RPC server procedure to be called.

❑ *RPC protocol version number.*

ArcView can be an RPC server while simultaneously making RPC requests. The server machine name is the host name. With the use of the host name, the RPC servers running on the host can be identified. The *rpcinfo* utility provides RPC information in the UNIX environment. In the following example, the registered RPC services on a machine named *washington* are displayed.

```
/usr/etc/rpcinfo -p washington
```

The *rpcinfo* utility returns the server identification number under the program column, along with the RPC version number for the program.

ArcView as RPC Client

ArcView as an RPC client can connect to an RPC server and request services. When making an RPC request, ArcView waits for the results before proceeding. If the time-out expires before the RPC server completes the request, the RPC is canceled. The default time-out is 20 seconds, but this can be changed. In the following example, an RPC client is established, a higher time-out value is set, and an RPC request is made.

14-03.AVE

```
' Connect to server on host named washington,
' with program ID of 0x4000001 and version number 1.
if (RPCClient.IsMachine("washington")) then
   if (RPCClient.HasServer ("washington", 0x40000001, 1))then
      anRPCClient = RPCClient.Make
      ("washington", 0x40000001, 1)
   else
     MsgBox.Error
     ("Invalid server ID for host washington",
     "")
     exit
   end
else
   MsgBox.Error
   ("Unable to connect to host washington",
   "")
   exit
end
'
' Verify that the connection was
' made and client is valid.
if (anRPCClient.HasError) then
   MsgBox.Error (anRPCClient.GetErrorID.AsString ++
   anRPCClient.GetErrorMsg, "")
   exit
else
   ' Set the time-out to at least 25 seconds.
   if (anRPCClient.GetTimeout < 25) then
     anRPCClient.SetTimeout (25)
   end
end
'
' Make a request and store the results in
' the respRPC string object.
ACommandString = "Is"
respRPC = anRPCClient.Execute(1, aCommandString, String)
```

An RPCClient object is always created when a Make request is sent. Therefore, the only way to verify the success of an RPCClient is to send the HasError request. Next, you can verify that the server machine is on the network and supports the server identification number. Use the IsMachine and HasServer requests to the RPCClient class to verify that the server is available before attempting to make an RPCClient object.

The Make request requires three parameters: host name, server ID, and version number. The Execute request sends a command to the RPC server. The first parameter for the Execute request is a procedure ID that alerts the server to the function expected from it.

ArcView as RPC Server

ArcView can also act as an RPC server. When performing as an RPC server, ArcView accepts only a script execution request. The procedure ID for executing a script is one (1).

The command line is the text of the Avenue script that is executed. For example, *av.GetProject.GetName* can be a command line. Predefined scripts can be run in the server by a command line such as *av.Run(scriptName, nil)*. ArcView always returns a string object to the RPCClient.

Only one instance of ArcView as an RPC server is viable. Use the following code segment to start an RPC server.

```
' Start an RPC server, while verifying
' that it is not currently running.
RPCServer.Stop
RPCServer.Start (0x4000001, 1)
```

The Solaris operating system recommends using server ID numbers between 0x40000001 and 0x5fffffff for customer-written applications. Verify whether server ID number ranges are recommended for other platforms.

An ArcView RPC client would use the following code segment to get the project name loaded into the ArcView server.

```
anRPCClient = RPCClient.Make
("washington", 0x40000001, 1)
anRPCClient.Execute (1,
"av.GetProject.GetName", String)
```

Using Dynamic Link Libraries

ArcView supports the use of Dynamic Link Libraries (DLL) in your application. You can write both Avenue scripts that call a DLL and DLLs that execute a script in ArcView. Both 32-bit and 16-bit DLLs can be used in a 16-bit operating system (OS) environment, but in a 32-bit OS environment ArcView supports only 32-bit DLLs.

You must be familiar with the function name, its returning data type, and parameters before you can access it in a DLL. The following steps access a function in DLL.

1. Load the DLL. In the statement below, a DLL named *mydll* is loaded.

```
myDLL.Make ("mydll".AsFileName)
```

2. Declare the function in the DLL. In the following code segment, *remove_blanks* is a function in *mydll* that returns a 32-bit integer and uses two strings in its parameter list.

```
myDLLFunc = DLLProc.Make (myDLL,"remove_blanks",
#DLLPROC_TYPE_INT32,
{#DLLPROC_TYPE_STR,#DLLPROC_TYPE_STR})
```

3. Call the function.

```
sourceString = "This string has blanks"
destString = String.MakeBuffer(sourceString.Count)
destCount = myDLLFunc.Call ({sourceString,
destString})
```

You can execute Avenue scripts from the DLL by including the *avexec32.h*, or *avexec16.h* header file in the code and linking with *avexec32.lib* or *avexec16.lib* library. ArcView installation places these files in *$AVHOME\lib* directory. Next, you can add the following statement to your DLL code.

```
resultString = AVExec(someAvenueScript)
```

The string variable *someAvenueScript* holds the script that you want ArcView to execute. You can use the av.Run statement to execute any Avenue script stored in the ArcView project.

Using Apple Scripts

ArcView can execute Apple Scripts on Macintosh systems using the AppleScript class. Apple Scripts can be used to launch other programs on your Macintosh or the network.

Create an Apple Script and store it ArcView as an Avenue script. Then use the following Avenue script to run it.

```
' Display a list of available Apple Scripts
' for selection. Assume that Apple Script names
' stored in the ArcView project start with
' AppleScript.
ASList = {}
for each aDoc in av.GetProject.GetDocs
  if (aDoc.Is(SEd)) then
    aDocName = aDoc.GetName
    if ((aDocName.Left(11))="AppleScript")then
      ASList.Add (aDoc)
    end
  end
end
ASDoc = MsgBox.List (ASList,
"Select an Apple Script to run.","")
if (ASDoc=nil) then
  exit
end
' Extract the AppleScript and replace
' new line characters with carriage
' returns.
ASCode = ASDoc.GetSource
ASCode = ASCode.Translate(NL, CR)
theAS = AppleScript.Make (ASCode)
if (theAS.HasError) then
  MsgBox.Error ("Error:"++
  theAS.GetErrorMsg,"")
```

```
else
  theAS.DoIt(nil)
end
```

ArcView for Macintosh supports Apple Events, which
make up the messaging system for Apple Scripts. Arc-
View supports inbound and outbound events, acting
as both client and server. Use the do-script Apple
Event to send an Apple Event to ArcView. The do-
script accepts one argument that should be an Avenue
script. Use the av.Run statement to execute scripts
already stored in ArcView project. You can use the
AppleEvent class to send Apple Events to other applications from ArcView.

Sample Application

ArcView is often used to develop applications that generate views from an index. Typically, an index view of map sheets is created and stored with a project. When a user selects one of the map sheets from the index, a new view based on the selected map sheet is created and themes are added. In some instances the themes are loaded automatically. At other times the user is prompted to select themes from a predefined set. The primary benefit of such applications is that the user is not required to know the name and directory structure of data files.

The sample application presented in this chapter was developed by the Baltimore County government in Maryland with the author's assistance. It has been updated for version 3.0 of ArcView and shortened for clarity. Eight Avenue scripts are listed and described below.

Data Query Application

The application is named Data Query and therefore all script names begin with DQ. This application has a new document type named Index, which is based on view documents. An Index document named *Grid View* is created with a theme named *index*. The index theme contains the map sheet polygons, and its feature table holds the map sheet numbers in a field named *plan_sheet*. Map sheet labels and a county boundary are also added to the Index document for visual reference. Grid View is stored with the project and becomes available when the project opens.

When the user selects a map sheet, she has the option of adding the selected map sheet to an existing view, or to a new view. A limit is set for the number of map sheets that can be added to a view. The view name is composed of the map sheet numbers.

The application has two customized controls: a pushbutton for the View GUI, and a new tool for the Index GUI. The pushbutton is used to remove map sheets from the views. Although views can be deleted, the last map sheet cannot be removed from a view because empty views are not permitted in the application. The tool is used with the Grid View to select a map sheet. Once a map sheet is loaded, a rectangle is drawn around the map sheet in the Grid View to highlight loaded sheets.

DQ.SelectMapSheet

This script is associated with the apply property of the customized tool for Index documents. It is executed when the user selects a map sheet from the index. The application establishes a global variable for the location of coverages, and the coverage directories start with the letter *t* followed by the sheet number. The global variable contains the letter *t*, but the sheet number is added as each map sheet is loaded.

 15-01.AVE

```
' ==========================================
' DQ.SelectMapSheet
' Apply the event of a tool on the grid index.
' Calls scripts to establish a new or
' existing view as the destination
' view, and then calls a script to load the map
' sheet in the destination view.
' ==========================================


' Set the home directory for the plans. Each
' directory starts with t followed by the
' grid number.
_planHomeDir = "d:/data/arcview/balt/t"
'

' Get the view and theme.
```

```
theview = av.GetActiveDoc
theDisplay = theView.GetDisplay
theTheme = theView.FindTheme("Index")
'
' The table for index theme lists the map sheet
' number in the plan_sheet directory.
theFTab = theTheme.GetFTab
theGridField = theFTab.FindField("Plan_sheet")
if (theGridField = nil) then
   MsgBox.Error ("Unable to find PLAN_SHEET field.","")
   exit
end

' Select the polygon that the user clicked on.
theSelection = theDisplay.ReturnUserPoint
theTheme.SelectByPoint(theSelection, #VTAB_SELTYPE_NEW)
theGrid = nil
for each aRec in theFTab.GetSelection
   theGrid = theFTab.ReturnValueString(theGridField, aRec)
end

if ((theGrid = nil) or (theGrid = "0")) then
   MsgBox.Info("No Grid Selected","")
   exit
end

' Establish the destination view.
destview = av.Run("DQ.EstablishView",theGrid)

if (destView = nil) then
   MsgBox.Warning ("Map sheet not loaded.","")
   exit
end

' Load the map sheet.
pList = {destView,theGrid}
worked = av.Run("DQ.LoadMapSheet",pList)

if (worked.not) then
```

```
      MsgBox.Warning ("Map sheet not loaded.","")
      exit
end


' Display the view document.
destView.GetWin.Open


' Update the grid view.
av.Run("DQ.UpdateGridindex","")
```

DQ.EstablishView

The view that receives the map sheet is identified in this script. Each view has a list of map sheet numbers as its object tag. The application uses the object tag to determine which map sheets are loaded in the view. This script also controls the number of map sheets that can be loaded into a view.

 CODE   15-02.AVE

```
' =======================================
' DQ.EstablishView
' SELF: map sheet number (string)
' RETURNS: nil or a view object
' This script checks to determine whether
' other views exist. If not, it creates a view
' and returns it. Otherwise, the user is asked
' to select an existing view or create a new view.
' =======================================


' Check the input.
if (SELF = nil or (SELF.Is(String).Not)) then
   MsgBox.Error("Internal Error - DQ.EstablishView, invalid SELF","")
   return (nil)
```

```
end

' Initialize variables.
viewsExists = false
newView = true
mapLimit = 2
tagList = {}
thegrid = SELF

' Are there other views?
doclist = av.GetProject.GetDocs
viewNameList = {"Use a new view"}
for each aDoc in docList
    if (aDoc.Is(View)) then
      if (aDoc.GetName <> "Grid View") then
        viewsExists = true
        viewNameList.Add (aDoc.GetName)
      end
    end
end

if (viewsExists) then
    ' Ask user to use an existing view or create a new one.
    destViewName = MsgBox.ListAsString
    (viewNameList,"Select the destination view.","")
    if (destViewName=nil) then
      return (nil)
    else
      if (destViewName = (viewNameList.Get(0))) then
        ' User has asked for a new view.
        newView = true
      else
        newView = false
      end
    end
end
```

```
' Perform what user has requested.
if (newView or (viewsExists.Not)) then
    destview = View.Make
    destView.SetName (SELF)
    tagList.Add (SELF)
    destView.SetObjectTag (taglist)
    return (destView)
else
    ' For existing views, verify that the user has not reached the limit and
    ' that the view is not a duplicate map sheet.
    destView = av.GetProject.FindDoc(destViewName)
    tagList = destView.GetObjectTag
    if (tagList = nil) then
      tagList = {}
    else
      for each msheet in tagList
        if (msheet = theGrid) then
          MsgBox.Error
          ("This map sheet already exists in the selected view.","")
          return (nil)
        end
      end
    end
    if (tagList.Count >= mapLimit) then
      msgBox.Error("View"++destViewName++
      "has reached the maximum number of"++
      "map sheets allowed.","")
      return (nil)
    end
    destView.SetName (destViewName+"/"+SELF)
    tagList.Add (theGrid)
    destView.SetObjectTag (tagList)
    return (destView)
end
return (nil)
```

DQ.LoadMapSheet

This script loads the themes from the selected map sheet.

 15-03.AVE

```
' =========================================
' DQ.LoadMapSheet
' SELF: List {View, map sheet number}
' RETURN: True (if successful), otherwise False.
' This script loads a predefined set of coverages and
' images into the given view. The themes are from the
' given map sheet.
' =========================================


' Verify input to the script.
if (SELF.is(List).Not) then
    MsgBox.Error("Internal error - DQ.LoadMapSheet, invalid SELF","")
    return (False)
else
    if (SELF.Get(0).Is(View).Not or
    (SELF.Get(1).Is(String).Not)) then
      MsgBox.Error
      ("Internal error - DQ.LoadMapSheet, invalid SELF members","")
      return (False)
    end
end


' Initialize variables.
destview = SELF.Get(0)
thegrid = SELF.Get(1).LCase
imageTab = av.GetProject.FindDoc("Image_cat")
if (imageTab = nil) then
```

```
        MsgBox.Warning ("Image catalog is missing.","")
        imageCat = nil
    else
        imageCat = imageTab.GetVtab
        imageField = imageCat.FindField ("Image")
    end
    somePlansMissing = false


    ' Set the coverage directory.
    planDir = _planHomeDir+theGrid+"/"
    orthoFile = nil
    if (imageCat <> nil) then
        for each aRec in imageCat
            imageFieldValue = imageCat.ReturnValueString(imageField,aRec)
            if (imageFieldValue.Contains(theGrid)) then
                orthofile = imageFieldValue
                break
            end
        end
    end


    ' Load the image first.
    if (orthoFile = nil) then
        MsgBox.Warning ("Image file not available.","")
    else
        aSrc = SrcName.Make (orthoFile)
        aTheme = ITheme.Make (ISrc.Make(aSrc))
        destView.AddTheme (aTheme)
        aTheme.SetVisible(True)
    end


    ' Building the view.
    aSrc = SrcName.Make(planDir+"bldg polygon")
    if (asrc = nil) then
        somePlansMissing = true
    else
```

```
   atheme = Theme.Make(aSrc)
   aTheme.SetName(theGrid++"BLDG")
   aSymb = aTheme.GetLegend.GetSymbols.Get(0)
   aSymb.SetColor (Color.GetBlack)
   destView.AddTheme (aTheme)
end


' Center line
aSrc = SrcName.Make(planDir+"cline arc")
if (asrc = nil) then
   somePlansMissing = true
else
   aTheme = Theme.Make(aSrc)
   aTheme.SetName(theGrid++"CLINE")
   aSymb = aTheme.GetLegend.GetSymbols.Get(0)
   aSymb.SetColor (Color.GetBlack)
   destView.AddTheme (aTheme)
end


if (somePlansMissing) then
   MsgBox.Warning ("Some of planimetric layers are not available.","")
end

'
' Finis.
return (true)
```

DQ.UpdateGridIndex

Every time a view or map sheet is added or removed, this script is called to ensure that rectangles on the index theme are synchronized with the loaded map sheets. A dictionary is tagged to the Grid View that maintains map sheets which have a surrounding rectangle on the index theme. This script identifies the loaded map sheets and then synchronizes with the index theme using the tagged dictionary.

 15-04.AVE

```
' ==========================================
' DQ.UpdateGridIndex
' Called from the events that may change the composition
' of the selected map sheets. This script updates the
' grid view by identifying map sheets that have been loaded.
' ==========================================


' Get the grid view and examine its tag object.
gridView = av.GetProject.FindDoc ("Grid View")
tagDict = gridView.GetObjectTag
if (tagDict = nil) then
    tagDict = Dictionary.Make (20)
    gridView.SetObjectTag (tagDict)
end


'Assemble a list of all loaded map sheets.
bigList = {}
for each aDoc in av.GetProject.GetDocs
    if (aDoc.Is(View)) then
        if (aDoc.GetName <> "Grid View") then
          aList = aDoc.GetObjectTag
```

```
        if (aList <> nil and (aList.Is(List))) then
          bigList = bigList.Merge(aList)
        end
      end
    end
end
bigList.RemoveDuplicates


' Add a rectangle to each available
' map sheet.
for each mSheet in bigList
    exists = tagDict.Get(mSheet)
    if (exists = nil) then
      av.Run("DQ.GridTurnOn",mSheet)
      tagDict.Add(mSheet,True)
    end
end
' Remove rectangles in which
' map sheets are not loaded.
dictList = tagDict.ReturnKeys
for each mSheet in dictList
    foundIt = bigList.Find(mSheet)
    if (foundIt < 0) then
        av.Run("DQ.GridTurnOff",mSheet)
        tagDict.Remove (msheet)
    end
end
```

DQ.RemoveMapSheet

This script is associated with the click event of the pushbutton that removes map sheets from a view. A list of loaded map sheets in the active view is presented to the user for selection. The application then removes the themes related to the selected map sheet. The user cannot remove the last map sheet from the view.

 15-05.AVE

```
' ========================================
' DQ.RemoveMapSheet
' Click event for a button on the view document
' GUI. Shows loaded map sheets based on the
' view's object tag. If the user selects to remove
' one, and it is not the last map sheet,
' it is removed.
' ========================================

' Get the view and examine its tag object.
theview = av.GetActiveDoc
tagList = theView.GetObjectTag
if (tagList = nil) then
    exit
end
if (tagList.Is(List).Not) then
    exit
end
if (tagList.Count = 0) then
    exit
end

mSheetCount = tagList.Count
```

```
' Display the map sheets.
msheet = MsgBox.ListAsString
(tagList,"Select a map sheet to remove.","")
if (mSheet = nil) then
    exit
end
if (mSheetCount <= 1) then
    MsgBox.Error ("The last map sheet cannot be deleted.","")
    exit
end

cutlist = {}

' Make everything inactive.
for each aTheme in theView.GetThemes
    aTheme.SetActive(false)
end

' Do not modify mSheet. It is needed to update taglist.
mLCSheet = mSheet.LCase

' Remove map sheet themes.
atheme = theView.FindTheme(mLCSheet++"BLDG")
if (atheme <> nil) then
    cutList.Add (aTheme)
end
aTheme = theView.FindTheme(mLCSheet++"CLINE")
if (aTheme <> nil) then
    cutList.Add (atheme)
end

' Remove image.
aTheme = theView.FindTheme(mLCSheet+".bil")
if (aTheme <> nil) then
    cutList.Add (aTheme)
end

' Remove all themes in the cutList.
```

```
for each aTheme in cutList
   theView.DeleteTheme(aTheme)
end


' Update tag and grid.
tagList.RemoveObj (mSheet)


' Update the views title.
theView.SetName (tagList.Get(0))


av.Run("DQ.UpdateGridIndex","")
```

DQ.GridTurnOff

This script removes a single rectangle from the index theme. Each rectangle is tagged with its sheet number. The sheet number is used to find the rectangle and remove it from the graphic list of the Grid View.

 15-06.AVE

```
'=========================================
'DQ.GridTurnOff
'SELF: The grid number
'This script finds the rectangle around a
'grid by its object tag (sheet number) and
'removes it.
'=========================================

'Initialize variable.
theGrid = SELF

' Get the graphic list.
gridView = av.GetProject.FindDoc("Grid View")
gridGL = gridView.GetGraphics

'Find the rectangle.
for each aShape in gridGL
   if (aShape.GetObjectTag = theGrid) then
     gridGL.RemoveGraphic(aShape)
     break
   end
end
```

DQ.GridTurnOn

This script places a single rectangle around a selected polygon on the index theme. Each rectangle is tagged with a sheet number so that it can be identified later.

 15-07.AVE

```
' =========================================
' DQ.GridTurnOn
' SELF: The grid number
' This script visually indicates which
' grid was selected. It expects one grid
' polygon to be selected. The map sheet number
' is tagged to the shape so that it
' can be found later.
' =========================================

' Get the extent of the selected sheet.
gridView = av.GetProject.FindDoc("Grid View")
gridGL = gridView.GetGraphics
indexTheme = gridView.FindTheme("Index")
aRect = indexTheme.GetSelectedExtent
'
' Add the rectangle to the index.
if (aRect.IsNull.Not) then
   aShape = GraphicShape.Make(aRect)
   aShape.GetSymbol.SetOLWidth (4)
   aShape.SetObjectTag(SELF)
   gridGL.Add (aShape)
end
```

DQ.DeleteDoc

Use this script to replace the click property of the Project/Delete menu item with a modified system script, Project.Delete. It updates the index if a view is deleted.

 15-08.AVE

```
' ==========================================
' DQ.DeleteDoc
' This is the modified Project.Delete system script.
' Replace the click property of the Project/Delete
' menu item with this script.
' It updates the grid view each time a view is deleted.
' ==========================================
theProject = av.GetProject
theDocs = theProject.GetSelectedDocs
doc_names = ""
for each d in theDocs
    if (doc_names = "") then
        doc_names = doc_names ++ "'" + d.GetName + "'"
    else
        doc_names = doc_names + "," ++ "'" + d.GetName + "'"
    end

    if(d.Is(View)) then
        editThm = d.GetEditableTheme
        if (editThm <> nil) then
            doSave = MsgBox.YesNoCancel("Save Edits to "+editThm.GetName+" in "+
            d.GetName+"?", "Stop Editing", true)
            if (doSave = nil) then
                return nil
            end
            if (editThm.StopEditing(doSave).Not) then
                MsgBox.Info("Unable to Save Edits to Theme "
```

```
                         + editThm.GetName +
                         ", please use the Save Edits As option", "")
              return nil
          else
              d.SetEditableTheme(NIL)
          end
       end
     end
     if (d.Is(Table)) then
       if (d.GetVTab.IsBeingEditedWithrecovery) then
         doSave = MsgBox.YesNoCancel("Save Edits to the table "+d.GetName+
         "?", "Stop Editing", true)
         if (doSave = nil) then
            return nil
         end
         if (d.GetVtab.StopEditingWithRecovery(doSave).Not) then
            MsgBox.Info("Unable to Save Edits to Table " + d.GetName +
                       ", please use the Save Edits As option", "")
            return nil
         end
       end
     end
   end
end
if (MsgBox.YesNo("Are you sure you want to delete " +
    doc_names + "?", "Delete", true)) then
    for each d in theDocs
      theProject.RemoveDoc(d)
    end
    ' The following line is added to the system script.
    av.Run("DQ.UpdateGridIndex","")
end
av.PurgeObjects
```

Avenue Class Hierarchy

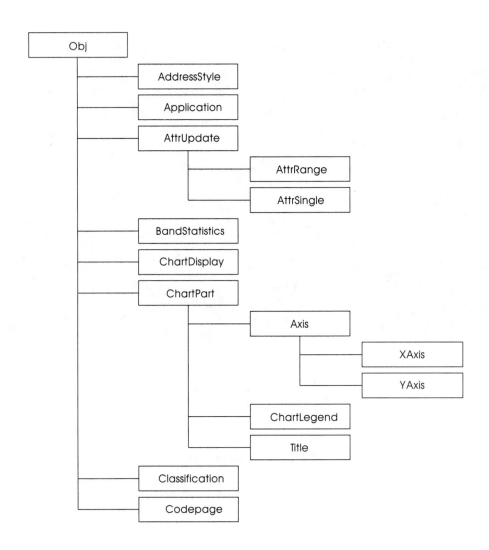

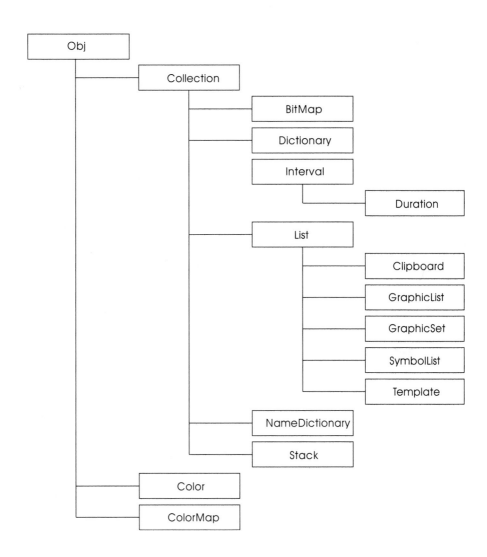

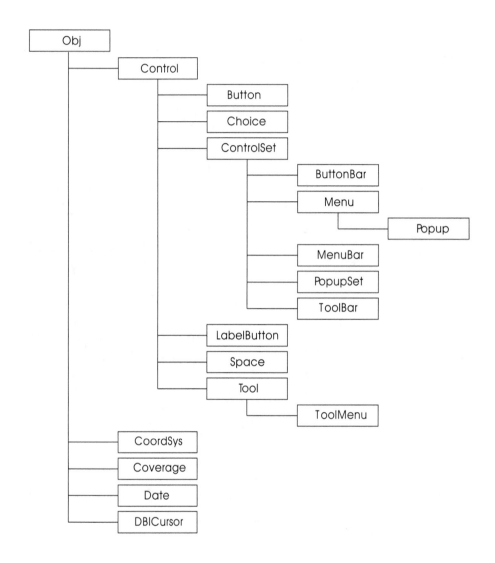

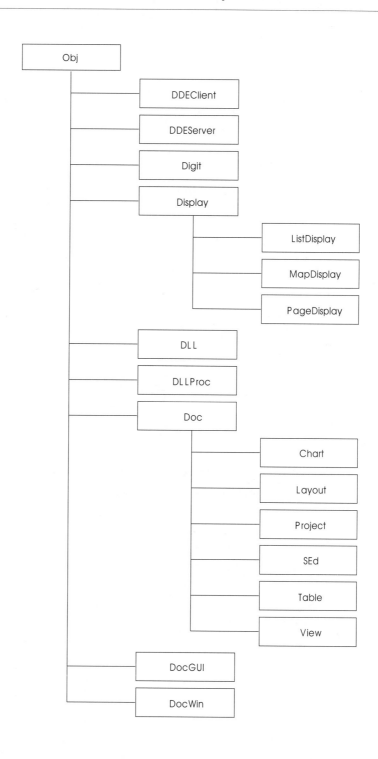

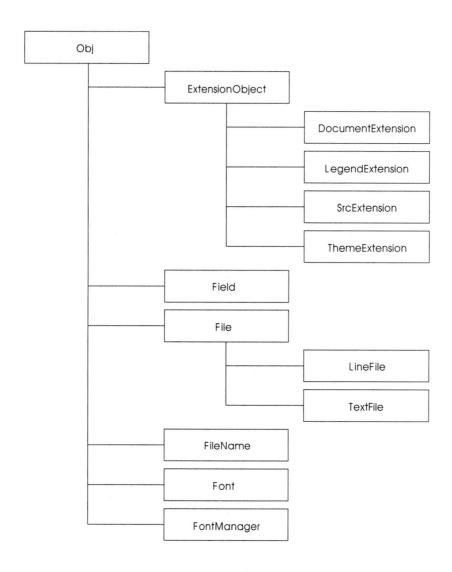

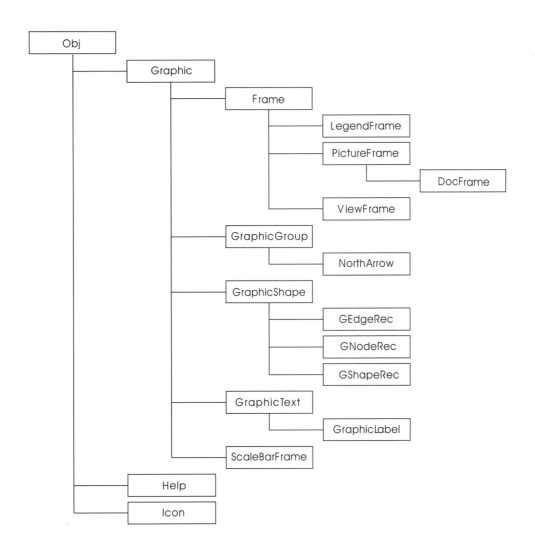

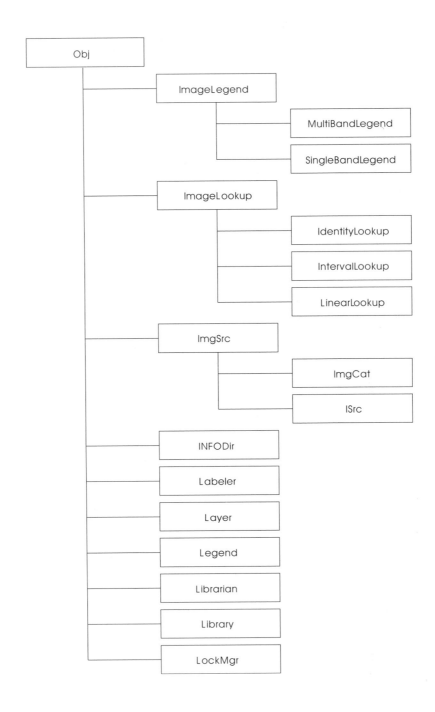

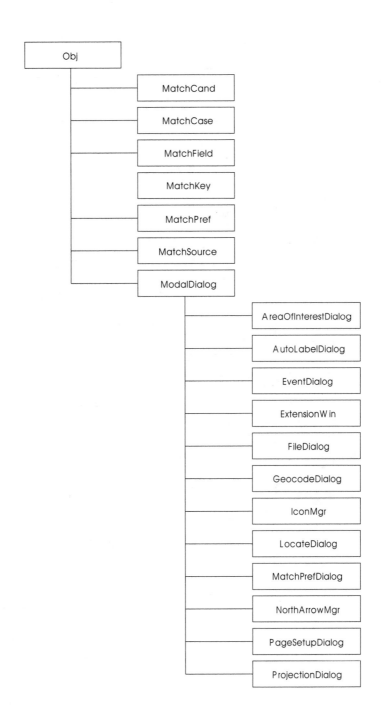

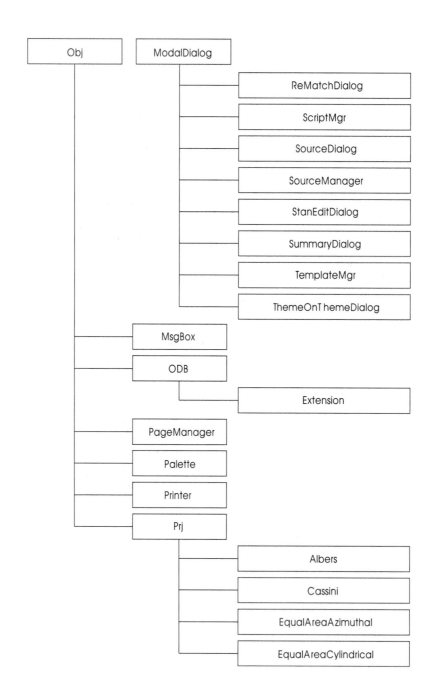

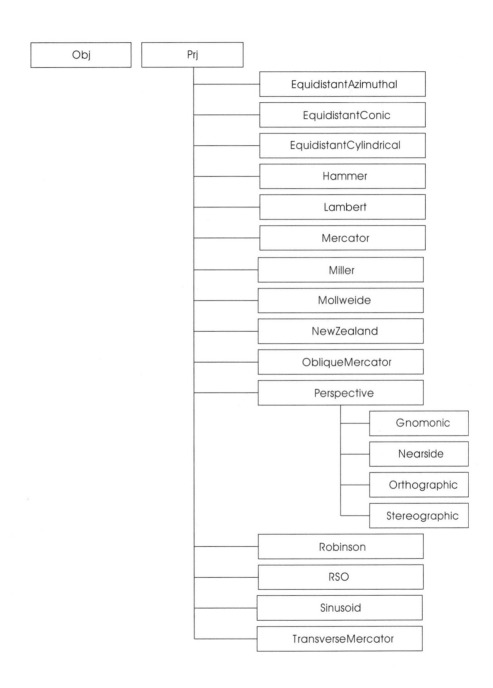

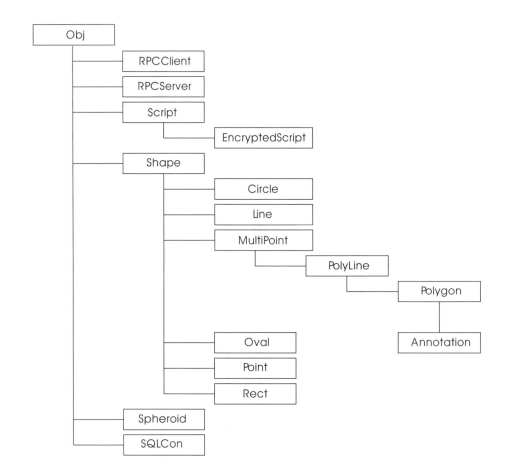

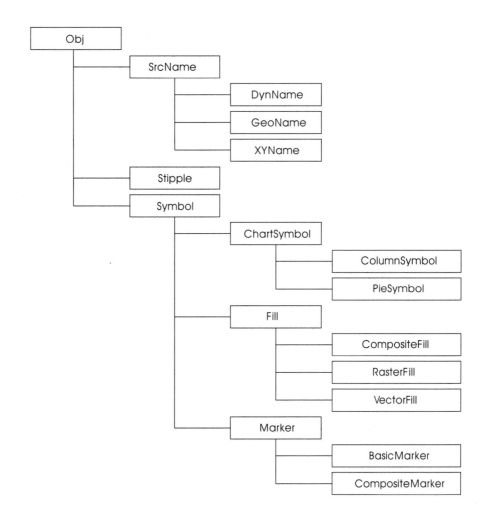

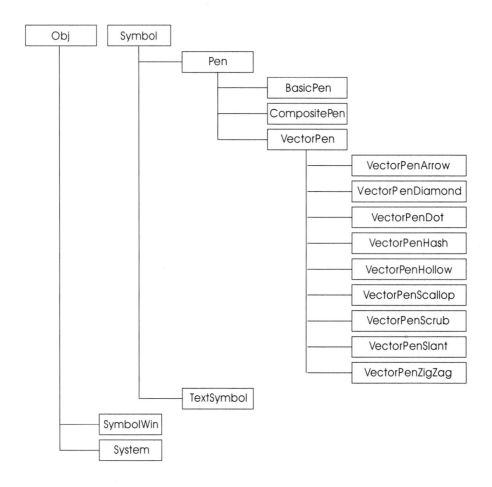

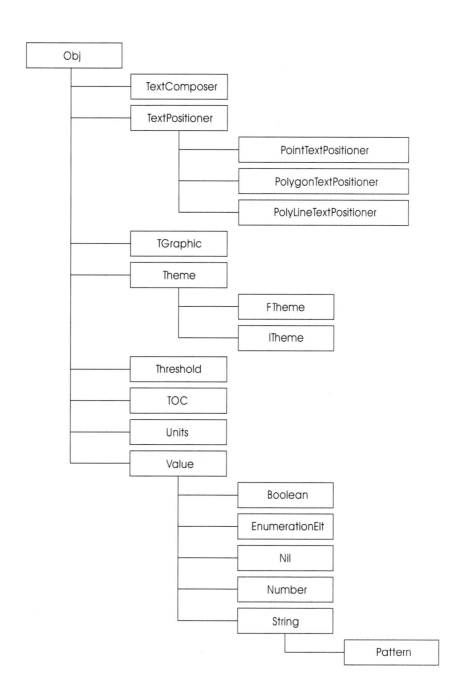

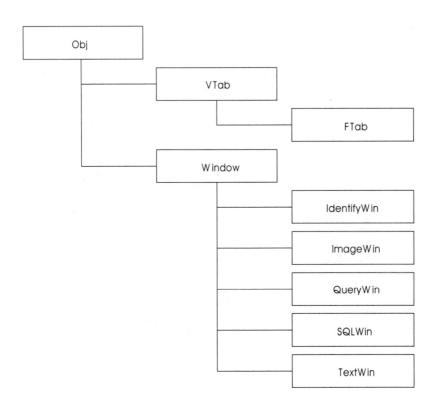

Avenue
Reserved Words

The following words are reserved for use by Avenue and should not be used as variable names.

AddressStyle	Clipboard	DynName
Albers	Codepage	Each
And	Collection	Else
Annotation	Color	Elseif
Application	ColorMap	EncryptedScript
AreaOfInterestDialog	ColumnSymbol	End
AttrRange	CompositeFill	EnumerationElt
AttrSingle	CompositeMarker	EqualAreaAzimuthal
AttrUpdate	CompositePen	EqualAreaCylindrical
AutoLabelDialog	Continue	EquidistantAzimuthal
Av	Control	EquidistantConic
Axis	ControlSet	EquidistantCylindrical
BandStatistics	CoordSys	EventDialog
BasicMarker	Coverage	Exit
BasicPen	Date	Extension
BitMap	DBICursor	ExtensionObject
Boolean	DDEClient	ExtensionWin
Break	DDEServer	False
Button	Dictionary	Field
ButtonBar	Digit	File
Cassini	Display	FileDialog
Chart	DLL	FileName
ChartDisplay	DLLProc	Fill
ChartLegend	Doc	Font
ChartPart	DocFrame	FontManager
ChartSymbol	DocGUI	For
Choice	DocumentExtension	Frame
Circle	DocWin	FTab
Classification	Duration	FTheme

GEdgeRec

GeocodeDialog

GeoName

GNodeRec

Gnomonic

Graphic

GraphicGroup

GraphicLabel

GraphicList

GraphicSet

GraphicShape

GraphicText

GShapeRec

Hammer

Help

Icon

IconMgr

IdentifyWin

IdentityLookup

If

ImageLegend

ImageLookup

ImageWin

ImgCat

ImgSrc

In

INFODir

Interval

IntervalLookup

ISrc

ITheme

LabelButton

Labeler

Lambert

Layer

Layout

Legend

LegendExtension

LegendFrame

Librarian

Library

Line

LinearLookup

LineFile

List

ListDisplay

LocateDialog

LockMgr

MapDisplay

Marker

MatchCand

MatchCase

MatchField

MatchKey

MatchPref

MatchPrefDialog

MatchSource

Menu

MenuBar

Mercator

Miller

ModalDialog

Mollweide

MsgBox

MultiBandLegend

MultiPoint

NameDictionary

Nearside

NewZealand

Nil

NorthArrow

NorthArrowMgr

Not

Number

ObliqueMercator

ODB

Or

Orthographic

Oval

PageDisplay

PageManager

PageSetupDialog

Palette

Pattern

Pen

Perspective

PictureFrame

PieSymbol

Point

PointTextPositioner

Polygon

PolygonTextPositioner

PolyLine

PolyLineTextPositioner

Popup

PopupSet

Printer

Prj

Project

ProjectionDialog

QueryWin

RasterFill

Rect

ReMatchDialog

Return

Robinson

RPCClient

RPCServer

RSO

ScaleBarFrame

Script

ScriptMgr

SEd

Shape

SingleBandLegend

Sinusoid

SourceDialog

SourceManager

Space

Spheroid

SQLCon

SQLWin

SrcExtension

SrcName

Stack

StanEditDialog

Stereographic

Stipple

String

SummaryDialog

Symbol

SymbolList

SymbolWin

System

Table

Template

TemplateMgr

TextComposer

TextFile

TextPositioner

TextSymbol

TextWin

TGraphic

Theme

ThemeExtension

ThemeOnThemeDialog

Then

Threshold

Title

TOC

Tool

ToolBar

ToolMenu

TransverseMercator

True

Units

Value

VectorFill

VectorPen

VectorPenArrow

VectorPenDiamond

VectorPenDot

VectorPenHash

VectorPenHollow

VectorPenScallop

VectorPenScrub

VectorPenSlant

VectorPenZigZag

View

ViewFrame

VTab

While

Window

XAxis

Xor

XYName

YAxis

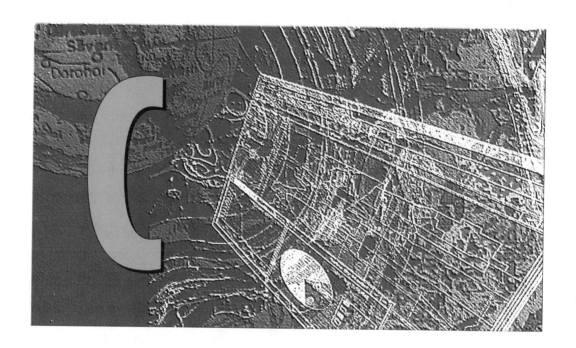

Changes in 3.0

Some class and object requests in ArcView 2.1 have changed in version 3.0. Update your scripts written in 2.1 if you have used any of the requests in the following list.

1. When an operation fails, the following requests return a null shape in version 3.0. When an operation failed in version 2.1 they returned nil. Examine the success of an operation with an IsNull request instead of comparison with nil.

   ❏ ReturnUserCircle

   ❏ ReturnUserLine

   ❏ ReturnUserPolygon

   ❏ ReturnUserPolyLine

   ❏ ReturnUserRect

2. Change the request GetExtent to ReturnExtent for instances of the following classes.

   ❏ FTheme

   ❏ ITheme

   ❏ Theme

3. Change SetReportUnits to SetDistanceUnits. Likewise, change GetReportUnits to GetDistanceUnits. These requests apply to Display class instances.

4. Change GetSysDefaults and GetUserDefaults to GetSysDefault and GetUserDefault. These requests apply to Application class instances.

5. The requests RampColors and RandomColors are no longer available for the objects of the Legend class. Use these requests with the symbol list of your legend.

6. The Default request is no longer valid for objects of the Legend class. Consider using the Single-Symbol request instead.

7. ArcView 3.0 no longer recognizes the enumeration element #UNITS_LINEAR_PRJMETERS.

8. Change the GetActive and SetActive requests for the project instance to GetSelectedDocs and Set-SelectedDocs.

9. Change the GetActiveClass request to GetSelectedGUI for the objects of the Project class.

10. Change the statement aDoc.GetPluralClassName to aDocGUI.GetTitle.

11. The GetFileName request for DLL class objects returns a file name object instead of a string in version 3.0.

12. Change the statement av.LoadLibrary or av.Unload-Library to System.LoadLibrary or System.Unload-Library.

13. Change the GetISrc request for ITheme class instances to GetImgSrc.

14. Names for several of the Prj subclasses have changed as listed below.

- ❏ Eqarazi to EqualAreaAzimuthal

- ❏ Eqarcyl to EqualAreaCylindrical

- ❏ Equiazi to EquidistantAzimuthal

- ❏ Equicon to EquidistantConic

- ❏ Equicyl to EquidistantCylindrical

- ❏ Oblmerc to ObliqueMercator

- ❏ Trnmerc to TransverseMercator

15. The name of the following subclasses of Perspective have changed.

- ❏ Ortho to Orthographic

- ❏ Stereo to Stereographic

16. Class ISrc is now a subclass of ImgSrc instead of Obj.

17. The argument list of the Contains request for the objects of Classification class has changed from (aValue, numDecimals) to (aValue).

18. Change the GetMaximum and GetMinimum requests of Classification instances to ReturnMaximum and ReturnMinimum.

19. Change the argument list of the MakeUsingDialog request of the Chart class from (aVTab) to (aVTab, aGUIName).

20. In ArcView 3.0, the Export request for the objects of Layout or View returns an instance of FileName.

21. Change aProject.MakeDocName(aClass) statements to aProject.MakeDocName(aDocGUI).

22. In ArcView 3.0, an aProject.Save statement returns a Boolean object.

23. ArcView 3.0 no longer recognizes the NextSelect request for objects of the Graphic class.

24. For objects of the Legend class, change GetField and SetField requests to GetFieldNames and SetFieldNames. ArcView 3.0 no longer recognizes GetMaxClassifications, SetMaxClassifications and IsSimple requests for instances of the Legend class. The statement aLegend.GetSymbols returns an instance of the SymbolList class.

25. Change the argument list for Interval or Quantile requests of legend objects to (aTheme, aString, numClasses). Change the argument list for aLegend.Unique to (aTheme, aString). Change the argument list for aLegend.Load to (aSavedLegend, loadType).

26. Change the aMatchSource.ReMatch statement to aMatchSource.BatchMatch.

27. The EventDialog.Show requires a view object as its argument in ArcView 3.0.

28. The statement MatchPrefDialog.Show (anMPref) returns a Boolean object in ArcView 3.0.

29. Change the argument list of ProjectionDialog.-Show to (aView, viewUnits).

30. The following requests in the Prj class have changed.

  ❐   GetEllipsoid to GetSpheroid

  ❐   IsNil to IsNull

  ❐   MakeNil to MakeNull

  ❐   SetEllipsoid to SetSpheroid

31. AddBreak and ClearBreak requests are no longer valid with objects of the Script class. Use these requests with the instances of the SEd class.

32. Change the argument list of anFTheme.SetSnapping to (allowSnapping).

33. Change the argument list of Window.Make to (libFile, libName, modName, resName, reload).

34. Change the argument list of anIdentifyWin.Add to (aClientObject, aKey, aName).

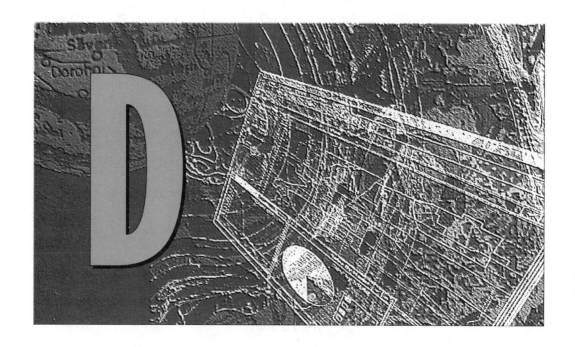

Avenue
Programming
Guidelines

Using programming guidelines or standards makes
Avenue scripts easier to understand and maintain.
Use the suggestions in this appendix to create a cod-

ing guideline or standard booklet for yourself or your organization. Update the booklet as you encounter topics that require guidelines or standards.

Script Name

Script names should have two parts separated by a dot. On the left side of the dot use the application name or an abbreviation of the application name. On the right side of the dot use a word or words that describe the script's functionality. Avoid using spaces or underscores in the name. Appearing below are examples from an application named Data Query.

❐ DQ.EstablishView

❐ DQ.GridTurnOn

❐ DQ.RemoveMapSheet

By starting each script name with the application name you will more easily locate scripts in the script manager. You should also make a habit of completing the property dialog box for each script.

Header

The header is the block of comment lines at the top of a script. Its purpose is to provide a quick description of a script and any additional information helpful in using or maintaining the script. The following items should be provided in a header.

❏ Script name (useful when script is written to a text file).

❏ Script description.

❏ Programmer names or initials.

❏ Date created.

❏ Date and purpose of last change.

❏ SELF object (if required).

❏ Returned object (if applicable).

If you are creating a library of scripts you can also add a line of keywords for an easier search.

Comment line

Comments within the code lines are crucial to understanding and maintaining the code. Do not repeat what the code shows; instead, explain the reason for the action. Use a blank line to attract readers' attention to the comment. If you are placing a comment in an indented block, indent the comment as well.

```
' Bad example:
for each aTheme in cutList
' delete themes
theView.DeleteTheme (aTheme)
end

  ' Good example:
```

```
    ' Delete themes listed in the cutList.
for each aTheme in cutList

    'theView object is set to the active view.
    theView.DeleteTheme (aTheme)
end
```

Variable Name

Follow Hungarian notation for name variables. Start the variable names with lowercase letters and avoid using special characters. If the name consists of more than one word capitalize the words after the first one. Some examples of variables using Hungarian notation follow.

```
worked
gridView
_gridView (global variable)
```

Consider using words or abbreviations that describe the class of your variable. For example, gridView is a variable pointing to a view. Consider adopting abbreviations for some of the longer classes. For instance, gridViewGL is better than gridViewGraphicList. Suggestions for abbreviations appear below.

❐ BB (button bar)

❐ BM (bit map)

❐ CM (color map)

❐ Dict (dictionary)

❐ Dpy (display)

❏ GL (graphic list)

❏ GS (graphic set)

❏ GTxt (graphic text)

❏ Img (image)

❏ Lgnd (legend)

❏ LgndF (legend frame)

❏ MB (menu bar)

❏ MDpy (map display)

❏ NDict (name dictionary)

❏ PDpy (page display)

❏ PF (picture frame)

❏ SBF (scale bar frame)

❏ SL (symbol list)

❏ TB (tool bar)

❏ VF (view frame)

Alignment

Indent the block of text inside a conditional or loop expression. If you break a line, align the following line with the first line, as shown in the following example.

```
if (gridView = nil) then
  MsgBox.Error ("Unable to find the grid view",
  "Error")
  exit
end
```

File Name

ArcView uses specific file extensions for different types of files. Below is a list of file extensions used by ArcView.

❑ apr (ArcView project)

❑ ave (Avenue script)

❑ avx (ArcView extension)

❑ dbf (DBase file)

❑ odb (Object database file)

❑ shp (ArcView shape file)

❑ txt (Text file)

Use the above file extensions when creating similar files.

Reserved Words

Start all classes and requests with a capital letter and capitalize the following words. Examples include XAxis, VectorFill, Make, and FindTheme.

Use all uppercase letters for the reserved word SELF.

Start special class instances with lower case letters, such as in *av* or *true*.

Begin the reserved words of Avenue language elements with lowercase letters (e.g., *if, then,* and *for each*).

Index

More OnWord Press Titles

Computing/Business

Lotus Notes for Web Workgroups
$34.95

Mapping with Microsoft Office
$29.95 Includes Disk

Geographic Information Systems (GIS)

GIS: A Visual Approach
$39.95

The GIS Book, 3E
$34.95

Exploring Spatial Analysis in GIS
$49.95

INSIDE MapInfo Professional
$49.95 Includes CD-ROM

Raster Imagery in Geographic
Information Systems
$59.95

INSIDE ArcView GIS, 2E
$39.95 Includes CD-ROM

INSIDE ArcView, 1E
$39.95 Includes CD-ROM

ArcView Exercise Book
$49.95 Includes CD-ROM

ArcView GIS/Avenue Developer's Guide,
2E
$49.95 Includes Disk

ArcView/Avenue Developer's Guide, 1E
$49.95

ArcView/Avenue Programmer's
Reference
$49.95

ArcView GIS/Avenue Programmer's
Guide, 2E
$49.95

101 ArcView/Avenue Scripts: The Disk
Disk $101.00

ARC/INFO Quick Reference
$24.95

MicroStation

INSIDE MicroStation 95, 4E
$39.95 Includes Disk

MicroStation 95 Exercise Book
$39.95 Includes Disk
Optional Instructor's Guide $14.95

MicroStation 95 Quick Reference
$24.95

MicroStation 95 Productivity Book
$49.95

Adventures in MicroStation 3D
$49.95 Includes CD-ROM

MicroStation for AutoCAD Users, 2E
$34.95

MicroStation Exercise Book 5.X
$34.95 Includes Disk
Optional Instructor's Guide $14.95

Build Cell for 5.X
Software $69.95

101 MDL Commands (5.X and 95)
Executable Disk $101.00
Source Disks (6) $259.95

Pro/ENGINEER and Pro/JR.

*Automating Design in Pro/ENGINEER
with Pro/PROGRAM*
$59.95 Includes CD-ROM

INSIDE Pro/ENGINEER, 3E
$49.95 Includes Disk

Pro/ENGINEER Exercise Book, 2E
$39.95 Includes Disk

Pro/ENGINEER Quick Reference, 2E
$24.95

Thinking Pro/ENGINEER
$49.95

Pro/ENGINEER Tips and Techniques
$59.95

INSIDE Pro/JR.
$49.95

Softdesk

INSIDE Softdesk Architectural
$49.95 Includes Disk

*Softdesk Architecture 1 Certified
Courseware*
$34.95 Includes CD-ROM

*Softdesk Architecture 2 Certified
Courseware*
$34.95 Includes CD-ROM

INSIDE Softdesk Civil
$49.95 Includes Disk

Softdesk Civil 1 Certified Courseware
$34.95 Includes CD-ROM

Softdesk Civil 2 Certified Courseware
$34.95 Includes CD-ROM

Interleaf

INSIDE Interleaf (v. 6)
$49.95 Includes Disk

Interleaf Quick Reference (v. 6)
$24.95

Interleaf Exercise Book (v. 5)
$39.95 Includes Disk

Interleaf Tips and Tricks (v. 5)
$49.95 Includes Disk

Adventurer's Guide to Interleaf LISP
$49.95 Includes Disk

Other CAD

Manager's Guide to Computer-Aided Engineering
$49.95

Fallingwater in 3D Studio
$39.95 Includes Disk

Windows NT

Windows NT for the Technical Professional
$39.95

SunSoft Solaris

SunSoft Solaris 2.* for Managers and Administrators
$34.95

SunSoft Solaris 2.* Quick Reference
$18.95

SunSoft Solaris 2.* User's Guide
$29.95 Includes Disk

Five Steps to SunSoft Solaris 2.*
$24.95 Includes Disk

SunSoft Solaris 2.* for Windows Users
$24.95

HP-UX

HP-UX User's Guide
$29.95

Five Steps to HP-UX
$24.95 Includes Disk

OnWord Press Distribution

End Users/User Groups/Corporate Sales

OnWord Press books are available worldwide to end users, user groups, and corporate accounts from local booksellers or from Softstore/CADNEWS Bookstore: call 1-800-CADNEWS (1-800-223-6397) or 505-474-5120; fax 505-474-5020; write to SoftStore, Inc., 2530 Camino Entrada, Santa Fe, NM 87505-4835, USA or e-mail orders@hmp.com. SoftStore, Inc., is a High Mountain Press Company.

Wholesale, Including Overseas Distribution

High Mountain Press distributes OnWord Press books internationally. For terms call 1-800-4-ONWORD (1-800-466-9673) or 505-474-5130; fax to 505-474-5030; e-mail orders@hmp.com; or write to High Mountain Press, 2530 Camino Entrada, Santa Fe, NM 87505-4835, USA.

On the Internet: http://www.hmp.com

OnWord Press, 2530 Camino Entrada, Santa Fe, NM 87505-4835 USA